PENNSYLVANIA'S OIL HERITAGE

PENNSYLVANIA'S

OIL

HERITAGE

Stories from the Headache Post

SEAN K. MILLER

THE
History
PRESS

Published by The History Press
Charleston, SC 29403
www.historypress.net

All illustrations are drawn by the author.

Cover design by Marshall Hudson.

First published 2008

ISBN 9781540234308

Library of Congress Cataloging-in-Publication Data

Miller, Sean, 1967-
Pennsylvania's oil heritage : stories from the headache post / Sean Miller.
p. cm.

1. Petroleum industry and trade--Pennsylvania--History. 2. Petroleum industry and
trade--Pennsylvania--Biography. I. Title.
HD9567.P4M55 2008
338.2'728209748--dc22
2008013254

CONTENTS

ACKNOWLEDGEMENTS

There are many people who have come together before me and around me who have allowed me to bring these tales to you. While it would be impossible to name everyone, I can and do name those individuals below, along with my reasons for doing so.

The men and women of the Oil Region. It was their work that produced the history and tales that have allowed me to tell my tales.

The historians and folklorists who have come before me. It was these people who preserved the history and folklore so that I could learn it when I grew old enough.

My parents, Clair and Beverly Miller. Without them, I would not be here to tell these tales.

Ally-Karen Miller. My sister, who has put up with hearing these tales told and told and told again over the years and always has some new insight to offer.

Kristi Ward. My student, my partner and my friend.

The Emerald Lady and her Noble Lord. They have been after me for as long as they have known me to write down my tales, and now they can say to me, "It's about time."

You, friend. For without you and your desire to read these tales and learn what has come before, I would not have an audience to whom to tell these tales.

INTRODUCTION

G reetings friend. I sure would appreciate it if you would come up on the porch of the thirst parlor here and join me for a spell. It's a hot sunny day, and spending the afternoon in the shade listening to a good tale or three would be a pleasant way to pass the time.

I suppose the proper and neighborly thing for me to do would be to introduce myself and explain why I'm here. All right, I'll do just that. To begin with, my name is Sean Miller and I hail from a very small town in northwestern Pennsylvania. It's called Dempseytown or Dempseystown, depending on which road sign you read, and I'd be mighty surprised if you've ever heard of it.

Still, it's where I grew up and where I hang my hat when I'm not on the road. Being a professional entertainer, I find myself on the road a lot. I fall into the category of entertainers called "variety." I have several skills—the first one I picked up is the reason I'm here. It was something I learned many years ago at my grandfather's knee. I didn't know it at the time, and I'm fairly certain that he didn't either, but he taught me how to be a storyteller. I've been told that I've gotten fairly good at it over the years, and over the years I've come to specialize in telling tales from the region where I grew up.

Yes, I know that I told you that Dempseytown is not famous, but it happens to lie in the middle of the Oil Region. In fact, one of the yearly school fieldtrips while I was growing up was to the Drake's Well Museum in Titusville, about fifteen miles away from my family's farm. So, from a very young age I was immersed in the folklore and history of the Oil Region. Then, when I got old enough to have a job, I went to work on an oil lease and learned even more about the business. Mostly, what I learned was that I'd rather tell you about it than do it. No, that's not entirely true. I enjoyed the time I spent in the oil patch, but there just ain't much of an oil patch

Sean K. Miller, author and folkteller.

left around here now and the thought of working in the hot deserts of the Southwest or Arabia does not appeal to me.

So, after I graduated, I took off and joined the U.S. Navy, where I learned a lot and received a medical discharge for my service. I'm not bitter about that. If I could, I'd still be in the U.S. Navy, but that was not to be, and after I received my discharge I put myself through college. Somewhere along the way I fell into being an entertainer. As I said, my grandfather had taught me the basics of storytelling and my time in the U.S. Navy helped me refine my craft even further. When I started out as a storyteller, I was telling what I would call the standard stories that everyone who treads this path tells. However, it wasn't long before I realized that I needed to set myself apart from everyone else. With that in mind, I started telling the folklore and folkhistory of Pennsylvania.

It was while I was working with the Venango Museum of Art, Science and Technology on its production of *Oil on the Brain* that I chanced upon a driller named Gib Morgan. The way I encountered him is the only way anyone can meet him now, through his words and the tales that he created. For you see, Gib died in 1901 and is long gone from this earth. However, his tales are still here, and through the work of people like Mody Boatright and other historians, I was able to meet him. When I did, I decided that I liked this oil well driller from long ago. Not only that, but here was a collection of stories and tales that would set me apart from all but one other storyteller out there. In case you're wondering, his name is William Stumpf and, like me, he lives in Pennsylvania's Oil Region.

It was over ten years ago that I started telling the folklore and the folkhistory of the Oil Region, and I have learned a lot about it through the years. There's still a lot to learn, but people know who I am and enjoy hearing me tell the tales. Storytelling is a large part of the shows I present with my partner. We also eat fire, make balloon sculptures and she dances. It makes for an enjoyable life that is never dull nor boring. As we are willing to travel, we get to see a lot of the country as well. I'm always willing to share my tales wherever we are, regardless of the show we are presenting.

I had been kicking around the idea of writing a book for a long time, but with everything else going on in my life, it was always on the back burner. It was just one of those projects that I planned on getting around to someday. Then, last November, I received an email from an editor at The History Press asking me if I'd be interested in putting some of my tales down on paper so that they could publish them in a book. As opportunity was knocking, I opened the door and let it walk in.

Introduction

The first request was for me to put my show down on paper and work that into a book. If you've ever been to more than one of my shows, friend, you'll understand how hard that would be. I doubt very highly if I've ever told the same tale the same way twice, let alone done two shows that were exactly the same. I know where I'm starting and I know where I have to go, but how I get there depends on the audience.

Let me explain it this way: basically, I have about sixty years of history, folklore and legends to choose to tell from. Now, if I get to a show and see that the audience is mainly adults, that's going to allow me to tell a different set of tales than if there were a lot of children in the audience. Also, I've learned over the years how to read an audience. Sometimes the people in the audience are more interested in the history of the region. At other times, they are more interested in the folklore or the people who lived "way back when." Of course, there are the shows where the people want a little bit of all of it to get a feel for the times. What all this means is that writing down my show is not only something that I would be hard-pressed to do, but also that I'd almost have to have a recording of a specific show to write from.

That would not make for a very interesting book, and the editors and I decided that it would be best to concentrate on just one segment of the collection of tales that I tell and see how you like them. As it's always best to start at the beginning, we further decided to start with the folkhistory tales and go on from there in later books. Therefore, the tales you will find in this book are all true, and I've done my best to make sure that the dates, locations and people are accurately represented.

I've written these tales as I tell them right now. If you've heard me tell these tales in the past, you may have heard them in a different form. If you hear me tell these tales in the future, I can guarantee that you won't hear the tales the way they are written here. It all has to do with the way I learn tales and the way I tell them. I learn the facts and figures and the roadmap of the tale. Then, when I get in front of an audience, we take the journey together. If we feel the need to spend a bit more time somewhere along the way, then so be it. If we come to a fork in the road and veer off into another tale, that's okay too, as we'll get there eventually and together.

While I've arranged the tales in this book in chronological order, there's nothing saying that they have to be read that way nor told that way. Each tale is just that: a tale about a piece of folkhistory from the Oil Region. It begins and winds its course until it gets to a point where it ends. Then, another tale begins, and while it's connected to those before it and those after it, it can stand by itself if need be.

At the end of this book, I've included a glossary of some of the words and phrases that were and are used in the Oil Region and industry. While it is by no means an exhaustive list, I did my best to give you a good idea of the way people talked back then. I've also included a timeline of some of the major events and happenings of the region. Finally, there is a list of other books that have been written about the region. Some of these are in print and some of them aren't, but many of them are written by historians and are much more scholarly than the book you're holding right now. Remember, I'm an entertainer and not a historian. Besides, I always break the word history down as "hi-story," emphasizing not the importance of the organized fact, but the true tales that brought us to where we are today.

Before we begin this journey of ours, I want to say a couple of words about the illustrations that you'll find throughout this book. They were all drawn by my hand in a style that I find hearkens back to the style of the illustrations that were used in the *Harper's Weekly* magazine during the late 1800s. I've based them off of photographs that were taken during that period, as well as sketches and illustrations that were done during that time. I wish I could tell you who took the photographs or drew the sketches from which I worked, but that information is lost to me. Often, all I had to work from was something that someone had torn out and saved because he liked it, and by the time it came into my hands, it was barely in one piece. As an artist myself, I would like to be able to say "he drew this" or "she took this picture" and thank the artists for doing so, but now, almost 150 years later, all I can do is tip my hat to all of them and thank them for recording the Oil Region as they saw it when it was new.

Now, friend, it is time to stand up and move out from in front of this thirst parlor and take a journey through Pennsylvania's Oil Region of way back when. Perhaps we can figure out how those events led to this modern world of ours. If after you've taken this journey with me you want to learn more about me and what I do, jump onto that newfangled thing they're calling the Internet and head over to www.greycloudstudios.com, where I've carved out a little space for myself.

KIER'S MEDICINAL PETROLEUM AND THE BIRTH OF AN INDUSTRY

If you were to start asking people where the petroleum industry began, more than likely you would get a couple of different answers, depending on where you were and who you asked. For the most part, you'd likely hear that Drake's Well at Titusville, Pennsylvania, was the place where the industry began. Unless you were talking to someone from the "Great State of Texas," in which case you'd be told that the industry began with the strike of the Lucas Gusher on Spindletop near Beaumont, Texas. The fact that this event happened fifty years after Drake's Well was drilled seems to mean nothing to the Texans. To them, Drake's Well was just a bunch of people playing around up in Pennsylvania before things really began in Texas.

What most people don't realize is that the petroleum industry had its start in a small town call Tarentum, which is just north of Pittsburgh, Pennsylvania, on the Allegheny River. It's true that no one ever drilled a well specifically for the purpose of striking petroleum prior to the work of Edwin L. Drake up in Titusville, Pennsylvania, but he wouldn't have been in the position he was in without the work that occurred in Tarentum during the fifteen years leading up to his strike. Most of this work was done by a certain Samuel M. Kier, a druggist from Pittsburgh.

Now, Kier's family owned several salt wells along the Allegheny River in the neighborhood of Tarentum. For you see, friend, salt production was big business and a lot of people were involved in the trade. The Allegheny–Ohio River Valley was a good place to drill down into the earth and find salt water. This is because of the way the rivers were formed during the last ice age. You may or may not be aware that they call Pittsburgh the city of the three

Samuel Kier, the first man to refine and sell petroleum commercially for medicine and illumination.

rivers. However, there is a fourth river that flows through the aquifer that lies under that valley basin. This river is not something that flows through a cavern and it doesn't look anything like the image you probably have in your mind right now. No sir, what you should be picturing is a long trough, cut into limestone rock and filled with limestone gravel of varying sizes. The water flows through this layer of gravel and over top of all of this lies several hundred feet of clay, sand and soil.

Some of the water that flowed through the ground had a high salt content, and when it was chanced upon by accident, some smart fellar figured out how to take this spoiled well of drinking water and make some money off of it. He figured that if he pumped up a bunch of this water and put it in a flat, shallow pond, the sun would dry up all the water. Once this was done, he could scrape up the salt and sell it in order to make himself rich. Well, friends, I can't say whether that fellar got rich, but it was such a good idea that a lot of people started doing it and an industry flourished.

The Kier family was one of the families who owned land along the Allegheny River and decided to go into business making salt. Well, friend, they did just that and from the money they made as salt well producers, young Samuel was able to become a chemist and druggist with a store in Pittsburgh.

All was not a bed of roses for these salt well drillers and producers, though, for there was a problem in the salt wells and it was a big problem. You see, from time to time, the salt wells would become polluted and ruined due to a smelly black liquid seeping into the well and coming up with the salt water. This liquid was called petroleum and it ruined many a salt well, mainly because no one was really sure what could be done with it. For the most part, when a well had petroleum in it to any great amount, the only thing that could be done was to pull the pipes and move on to the next well.

Why even good salt wells were not free of this petroleum problem. No sir, they had a similar but not salt well destroying problem to deal with. You have to understand that the petroleum in Pennsylvania is the best in the world. The reason for this has to do with all of the stuff that can be extracted from Pennsylvania Grade A Crude Oil. Of course, this does make it more expensive to refine and sometimes not worth the bother to pump out of the ground, but it was all of this extra stuff that was causing the problems in the good salt wells. For you see, one of the main items in all that stuff was a variety of paraffins. Yes sir, good old wax, and several forms of it, too. So what would happen was that these waxes would get into the pumping valves and start to build up. The drillers called this build-up petroleum butter, and while it was a pain to deal with, it was useful in other ways. For instance, it made a hair wax, if you could stomach the smell. Then, too, it seemed to help in the healing of burns and scrapes. So, it was not totally a bad thing; it just slowed down the works when it had to be scraped off of the pumping valves.

Well, Kier knew about this petroleum butter and more than likely kept a jar of it in his store. I'm speculating here, but I'd be willing to guess that he used it on someone or offered it to someone to use for treating a scrape or a burn and that someone liked it so much that he asked Kier where he could buy some. As I said, I'm not sure that that event ever happened, but what I am sure of is this: sometime around 1846, Kier started buying up all the petroleum butter he could get his hands on and then pressing it out into small, matchbox-sized cubes. He hand packaged these cubes in small boxes and sold them as a remedy for burns, scalds and bruises. His petroleum butter sold so well that he was regularly making shipments to the gold fields in California during the Gold Rush of 1849.

Now, the way I figure it was that the success he had with Petroleum Butter, along with the local tales of members of the Seneca Nation using liquid petroleum as a medicine, got him to thinking about the petroleum that often ruined salt wells. I'd be willing to bet that the first couple of production runs of "Kier's Medicinal Petroleum" were nothing more than trials to see if he had something salable. Well, friend, not only did he have something salable, but he also had the proverbial tiger by the tail. It wasn't too long before he had agents traveling about in wagons that were painted on one side with a salt well and on the other with a representation of the good Samaritan helping the ailing Hebrew with a bottle of his medicinal petroleum. His agents were so good at what they did that soon Kier was selling two barrels of oil a day at $0.50 a pint, which works out to be a gross profit of right around $336.00 a day.

If only Kier's Medicinal Petroleum had lived up to all its hype. An advertisement from the day included in John McLaurin's book, *Sketches in Crude Oil*, reads as follows:

Kier's Petroleum, or Rock Oil, Celebrated for its Wonderful Curative Powers. A Natural Remedy! Procured from a Well in Allegany Co., Pa., Four Hundred Feet below the Earth's Surface. Put up and Sold by Samuel M. Kier, 363 Liberty Street, Pittsburgh, Pa.

> *The healthful balm, from Nature's secret spring,*
> *The bloom of health and life to man will bring;*
> *As from her depths this magic liquid flows,*
> *To calm our sufferings and assuage our woes.*

The Petroleum has been fully tested! It was placed before the public as A REMEDY OF WONDERFUL EFFICACY. Every one not

One of the first methods for shipping petroleum from the well to the refinery.

acquainted with its virtues doubted its healing qualities. The cry of humbug was raised against it. It had some friends—those who were cured through its wonderful agency. Those who spoke in its favor. The lame through its instrumentality were made to walk—the blind to see. Those who had suffered for years under the torturing pains of *RHEUMATISM, GOUT, AND NEURALGIA* were restored to health and usefulness. Several who were blind were made to see. If you still have doubts, go and ask those who have been cured! * * * We have the witnesses, crowds of them, who will testify in terms stronger than we can write them to the efficacy of this remedy; cases abandoned by physicians of unquestionable celebrity have been made *to exclaim, "THIS IS THE MOST WONDERFUL REMEDY YET DISCOVERED!" * * * Its transcendent power to heal MUST and WILL become known and appreciated. * * * The Petroleum is a Natural Remedy; it is put up as it flows from the bosom of the earth, without anything being added to or taken from it. It gets its ingredients from the beds of substances which it passes over in its secret channel. They are blended together in such a way as to defy all human competition. * * * Petroleum will continue to be used and applied as a Remedy as long as man continues to be afflicted with disease. Its discovery is a new era in medicine.*

Now friend, Kier was doing quite well with his store in Pittsburgh, the Petroleum Butter and his Medicinal Petroleum, but like most men of the age, he was a tinkerer and could not leave well enough alone. It's a good thing too, for what he began to tinker with was the fact that petroleum burns and gives off light. At the time, most light was generated from wood or coal fires and lamps using whale oil. Truth be told, the world was beginning to run out of whales as they were being killed off faster than they could reproduce. So, Kier began to tinker. Raw petroleum has two major drawbacks when burned, one being the stench and the other being the smoke. No one in his right mind would tell you which one was worse and neither will I, but it was to these twin problems that Kier began to turn his mind.

Friend, it was on a trip to Philadelphia, Pennsylvania, that he got his answer in a roundabout way. He was discussing the problem with another chemist while he was in Philadelphia on business somewhere around 1849 or so, and this chemist, whose name is lost to history, suggested that Kier try refining the petroleum to see if that helped with things. The only problem was that he neglected to tell Kier how to go about doing so nor what equipment he would need.

Yet, as I've said before, Kier was a bright man and soon he had put together an apparatus consisting of a cover kettle and a worm. No, not the kind of worm that crawls on the ground. The worm that Kier used was a large spiral of tubing, which was attached to the top of the cover on the kettle and was used to let the vapors being boiled off of the petroleum condense back into liquid form. Needless to say, his first attempts did not produce the desired results.

It took him a while, but Kier learned how to control his flame and to run the petroleum twice through his still in order to get a liquid that had the color and appearance of apple cider. It still had an odor to it, but it burned well enough and produced only a slight amount of smoke. Kier had solved half of his problem and now he turned to the other half: coming up with something that would burn this new liquid, which he took to calling Carbon Oil. He began with a camphene lamp and had to make only slight changes to it in order for it to burn his Carbon Oil without producing any smoke.

After some additional improvements, including the addition of a Virna Burner and some further work on his distilling process, Kier was able to take both to market where both sold readily. In fact, he did so well that in 1854 he built and opened the world's first oil refinery at the intersection of Canal Street and Seventh Street in downtown Pittsburgh. While it only had a five-barrel still, it was nonetheless the first of its kind, and from it Kier shipped his Carbon Oil around the world.

As is often the case, history passed over Samuel Kier in favor of another, but he lived a good life. Now you know the truth about how the industry truly began—or, at least, one version of it.

DRAKE'S WELL

THE FIRST WELL DRILLED FOR THE COMMERCIAL PRODUCTION OF PETROLEUM

While Kier was going about his labors and marketing down in Pittsburgh, Pennsylvania, other people in other regions were starting to look at petroleum as a valuable commodity. One of these locations just happened to be on the Willard Farm, which lay about two miles south of Titusville, Pennsylvania, on Oil Creek.

At the time, and we're talking about the early 1850s here, the Willard Farm was in the possession of the firm of Brewer, Watson & Co. The owners of the firm were Ebenezer Brewer, James Rynd, Jonathon Watson, Rexford Pierce and Elijah Newberry, and they were lumbermen by trade. It was on an island in Oil Creek, a couple of rods south of the boundary between Crawford and Venango Counties, where Brewer, Watson & Co. had set up one of its sawmills. There was also a rusty-looking spring on this island through which petroleum bubbled to the surface to be scraped off and collected. The men at the sawmill used this petroleum to lubricate their saws and burned it in rude, smoky, chimneyless lamps for illumination.

Seeing that there might be some potential in this black, smelly stuff, the firm of Brewer, Watson & Co. signed an agreement with J.D. Angier of Titusville to further develop and maintain the spring as a source of petroleum. This document, the world's first oil lease, was signed on July 4, 1853, and it laid out what J.D. Angier was to do. It also stated that the profits, if any, would be divided equally between Angier and Brewer, Watson & Co. for the term of five years from the signing of the lease.

One of the first things that the firm did was to send a sample of the petroleum to one Professor Benjamin J. Silliman Jr., a chemist at Yale College in Dartmouth, New Hampshire, for a complete analysis. Professor Silliman

Agreed this fourth day of July A.D. 1853, with J.D. Angier of Cherrytree township, in the county of Venango, Pa. That he shall repair up and keep in order the old oil springs on land in said Cherrytree township, or dig and make new springs and the expenses to be deducted out of the proceeds of the oil, and the balance, if any, to be equally divided, the one half to J.D. Angier and the other half to Brewer, Watson & Co., for the full term of five years from this date. If profitable.

J.D. Angier

Ebenezer Brewer for

Brewer, Watson & Co.

The first written document looking into a mechanical development of petroleum production. This document was signed on July 4, 1853, by J.D. Angier of Cherrytree Township and the firm of Brewer, Watson & Co.

completed his report early in the spring of 1855, and while it's a dry, boring read, it provided the information that was needed for later investors. Among his many conclusions was that this liquid could be completely refined and there would be very little wasted at the end of the process.

Well, sir, while Professor Silliman was doing his work in far-off New England, J.D. Angier got down to business. He began by digging trenches that connected the spring to a shallow basin. From there, the water was pumped into a series of shallow troughs that sloped slowly to the ground. A series of small skimmers was set up in the troughs to skim the oil flowing by on top of the water so that it could be collected. By this means, he was able to gather three to four gallons a day. Unfortunately, this system did not turn a profit and so he stopped using it.

It was during the summer of 1854 that Dr. Francis Brewer, the son of Ebenezer Brewer, one of the owners of Brewer, Watson & Co., was visiting some friends in Hanover, New Hampshire, and stopped off at his alma mater, Dartmouth College, to see one of his old instructors, Professor Crosby. In his pocket, he carried a bottle of petroleum; a gift for the professor, who showed it to another graduate of his. That student was George Bissell and he was practicing law in New York City with Jonathon Eveleth at the time. It was the idea of Crosby's son to head out to Pennsylvania and give the spring a once-over. If things looked good, he wished to organize a company with a quarter of a million dollars in capital to purchase the lands around the spring and set up such machinery as might be needed to collect the petroleum on a commercial basis.

Well, friend, I'll tell you, there were all manner of misunderstandings and complications before everything was set into motion, but in the end Crosby, Bissell and Eveleth filed articles of incorporation for the Pennsylvania Rock Oil Company in both New York City and Albany, New York, on December 30, 1854. In January 1855, the Pennsylvania Rock Oil Company purchased the 150 acres of land that encompassed the island on which the sawmill and the spring were found and lands on either side of Oil Creek. However, almost as soon as they started, they had to reorganize in Connecticut in order to get out from under the law in New York that made each shareholder liable for the debts of the company.

At first, it looked like things were finally going to happen, but trouble reared its ugly head again. This was due to the fact that there were more people in charge, or who thought they were in charge, than people trying to get things done. Now, you have to understand that they were still trying to use skimmers in a profitable manner and failing miserably.

In fact, friend, things went along like that until the summer of 1856. It was a rainy day in New York City and the rain had forced George Bissell to take refuge beneath the awning of a drugstore on Broadway. While he was standing there, he saw a $400 bill on display in the window. This caused him to take a second look, and on further examination, he realized that it was a show bill for Kier's Medicinal Petroleum and there were a couple of bottles also on display in the window. Well, friend, the show bill had an illustration of a salt well on the top of it, and the number four hundred was not a dollar amount but a depth amount. For you see, that was the depth of the main salt well from which Kier got his petroleum.

Bissell stepped inside the drugstore and asked the druggist if he could have a closer look at the show bill. Much to his surprise, the druggist told him to take it along with him. Well, Bissell took that show bill back to his office and started studying it. As he did, he slowly came to the idea that if they were to make a go of things in Titusville, they might want to look into drilling for the source of the oil that was bubbling up through the spring. His next problem was to convince his business partner, Jonathon Eveleth, that this was the right path to follow. The main reason he had to do this was that he and Eveleth held the controlling interest in the Pennsylvania Rock Oil Company.

William "Uncle Billy" Smith, the first man to drill a well for the purpose of producing petroleum.

That afternoon, Eveleth jumped at the idea. Their first thought was to head off to Titusville on the next train and do the drilling themselves. Upon further reflection, they realized that this was not the thing to do, for not only did they have no idea how to bore a well, but it would also mean the end of their legal practice. At this point, they turned to James M. Townsend of New Haven, Connecticut, who was employed as president of the Pennsylvania Rock Oil Company. It was his idea to hire Edwin L. Drake to travel to Titusville and report on what he found there. Drake was instructed to travel first to Syracuse, New York, then to Erie, Pennsylvania, and from there to Titusville, returning by way of Tarentum and Pittsburgh.

Townsend had hired Drake for a couple of reasons. First, he was known by the principals of the company, having once been a stockholder. Second, he was a friend and had just gotten over a serious illness, and Townsend thought the change of air and scene would do Drake some good. Finally, there was the fact that Drake had retired from the New York & New Haven Railroad where he had been a conductor. Because of this, he had a pass that allowed him to travel for free on any railroad in the country. Drake set off on his journey in December 1857.

It was around the middle of the month when he reached Syracuse and stopped there to view the salt wells with an eye to adapting them to oil production. At Erie, he got off the train and had to take a mail wagon to Titusville. While there, he stayed at the American Hotel and spent his time examining the spring and the lands around the spring. Having done that, he traveled on to Tarentum to examine the salt wells there. He then went on to Pittsburgh to conduct some legal business for the Pennsylvania Rock Oil Company, arriving back in New Haven for the December 30 director's board meeting.

Things went quickly after that, and on March 23, 1858, the Seneca Oil Company was formed as a working partnership between the principals and Drake, with Drake being the president and owning a fortieth of the stock. Drake was then given $1,000 operating money and a salary of $1,000 a year. So it was that he moved himself and his family back to the American Hotel in Titusville, arriving there in early May.

It was either during this move or his trip the prior winter that Drake was first given the title of colonel. He was given the title by Townsend in the belief that the local populace of Titusville would be more willing to grant him respect and honor than if he did not have any such title. The legend is that Drake first learned of this fact when he checked into the American Hotel, only to find a letter addressed to Col. Edwin L. Drake.

Drake's first step was to rebuild the system of troughs and skimmers that Angier had set up five years before. By the end of June, he had adjusted the system enough so that he was producing ten gallons of petroleum a day. While he shipped his first two barrels back to New Haven in August, he was still having a hard time finding anyone to drill an artesian well for him. Also, in a further setback, the steam engine that he had bargained for never showed up, nor did the driller that Dr. Brewer had hired in Pittsburgh. All this was taken into account and the operations for the year were suspended.

Drake himself went to Tarentum in February 1859 and hired a driller. When this man failed to appear on time, Drake headed back to Tarentum. While there, he was told by F.N. Humes that the best tools were made by William "Uncle Billy" Smith, who might be available for the job. Friend,

The first Drake's Well. This rig burned on October 7, 1859, and was rebuilt with a rig resembling the replica that now stands at the wellhead.

Smith was not only available, but he accepted the job and on May 20, he arrived in Titusville with his two sons to help him drill the well for Drake. By this point, a pump house had been framed and a derrick and boardinghouse had been built on site.

At this point, friend, you need to understand how one drilled an artesian well. To begin with, one dug a timber-braced pit down until bedrock was found. Once one caught rock, as the phrase went, the drilling of the hole could commence. Well, sir, on that island, with a spring bubbling up through it like it did, it was durn near impossible to dig a pit, let alone have it stay open until you could line it with timber. Workers fought that problem for some time until Drake had the idea to drive a pipe down through the ground until they reached the bedrock and could begin to drill. This they did, and they reached the bedrock at a depth of thirty-six feet. They started to drill on August 14.

The drilling progressed at a rate of three feet a day, and the workers rested at night and on Sunday. At this rate, Drake figured that they would reach the four-hundred-foot depth of the Tarentum wells by Christmastime. However, the people with the money back in New England were not being very forthcoming due to the repeated delays and misunderstandings that had been more or less constant since the company was formed. Drake had been able to locally secure $600 to help him weather the current crisis.

Well, friend, that was how things stood on the afternoon of August 28, 1859. Uncle Billy had reached a depth of sixty-nine feet in coarse sand and he and his boys decided to lay off until Monday morning. However, just as they were about to quit, the drill string dropped another six inches into a crevice. They were not too concerned about this, as it happened often when drilling a salt well. Thus, the tools were drawn out and the well was allowed to rest until Monday.

On Sunday, August 29, 1859, Uncle Billy was heading out with his sons to do a bit of fishing when he stopped by the well to make sure everything was still okay. As he looked down the pipe, he was surprised to find that it was full of liquid to within eight to ten feet of the surface. Grabbing a piece of tin spouting that was laying nearby, he plugged one end and tied a line to the other. Then he lowered it down into the liquid, and when he drew it up, he found it to be petroleum.

He spent the rest of the day running that piece of tin spout down the well and pulling it out again, each time full of petroleum. It was a stranger who was out for a stroll that carried the news into Titusville proper, for Uncle Billy was afraid to leave the wellhead. By the time Drake arrived bright and early the next morning, Uncle Billy and his boys had drawn out three full barrels of the precious liquid. By noon, they had adjusted the pumping apparatus and the well commenced to pumping twenty barrels a day.

The ironic thing is that the mail on Monday brought a letter addressed to Col. Drake from the principals in New Haven informing him that he should suspend all further activity, liquidate the assets and return to New England. For a moment, just imagine what went through the principals' minds when they heard of the strike and realized how close they had come to missing the boat on the whole venture. It was just their good fortune that the mail was not faster and they became known as the people who drilled the first well in history specifically for the purpose of producing petroleum.

THE RUSH IS ON

THE FIRST SEVERAL MONTHS OF THE PETROLEUM INDUSTRY

After the end of August 1859, life along Oil Creek in Venango County, Pennsylvania, moved at a pace never before seen in that region of the world. Up until then, it had been a region of farmers and lumbermen whose lives moved at nature's pace, slow and steady. However, the news of Drake's Strike on August 29 spread through the valley like wildfire and launched an industry that changed the very way the world lived.

The first thing to happen was that crowds started to gather at the wellhead in order to take a look at this wonder of human ingenuity. Uncle Billy Smith wrote to his former employer, a fellar named Peterson, and told him to "come quick as there's oceans of oil to be had." Upon seeing the well for himself, Jonathon Watson, of Titusville, jumped on his horse and rode off down the creek at a gallop in order to secure the lease on the McClintock Farm. His reasoning was that it was already proved to have petroleum on it because several years before Nathanael Cary had constructed a timber crib and ran an oil-dipping operation from it.

Within days, Henry Potter of Titusville had tied up all the lands for miles along Oil Creek, hoping to secure the interest of some New York City investor. At the same time, William Barnsdale had secured the lease of the farm just north of the Willard Farm where Drake's Well was located. Finally, George H. Bissell bought up all of the Pennsylvania Rock Oil stock he could lay his hands on and was at the wellhead by the end of the week. Then, he began leasing farm after farm along Oil Creek and the Allegheny River, not paying any heed to what the surface indications were or what anybody said about what an idiot he was being for doing so.

A couple of stout fellows kicking down a well with a spring-pole rig.

Titusville itself grew up overnight. It went from a hamlet to a borough to a city of over fifteen thousand inhabitants. Jonathan Titus had first cleared the land to make himself a farm, and he lived to see his farm become "The Queen City." It was noted for its tasteful residences, excellent schools and active population. Many men settled there for the sake of their families. They paved the streets and planted shade trees along them. For a time, it became the center of the petroleum industry.

The sad part of this early growth was the narrow view Col. Drake took of things. He told many a man that he had "tapped the mine" and that no paying territory would be found outside of the land leased by his company. Therefore, he pumped his well and secured not one foot of ground for himself. Had he but known, he could have gone out that first week and secured for himself all the land he wanted. It would not have cost him a thing; a percentage of the royalties was often all that any farmer was interested in receiving in exchange for signing a lease. Even though his friends all encouraged him to get into the game, he rejected their counsel and didn't realize his mistake until it was too late for his name to mean anything to anyone.

The second well to be drilled was started a stone's throw to the north of Drake's Well by William Barnsdall, Boone Meade and Henry Rouse. It came in November 1869 at a depth of eighty feet. However, the output had an indigo tinge and the well yielded only five barrels of oil during the entire three days it was pumped. So, it was no surprise to anyone that operations were ceased for a couple of weeks. Drilling started again in December and continued until February, when a depth of 160 feet was reached. It was on February 19, 1860, that the tubing was put in and the well began producing fifty barrels a day.

In short order, The Crossely Well was finished on March 4 and flowed at sixty barrels a day. At that point, friend, the oil started coming in so fast that the local dealers just couldn't keep up with it. On June 1, fifty-six thousand barrels from The Barnsdall Well were sold in New York City for $17,000. However, the most amazing thing is that three of the wells that were started in 1859 were drilled by the spring-pole method.

Just so you know, friend, the spring-pole method works like this: first, you have to find a likely spot for there to be oil. This can be done by any number of methods, but generally you just have to make your pick and take your chances. After you've secured your location, you then go out into the woods and find a good sturdy tree of ash or hickory that has some flex to it. It should be anywhere from twelve to twenty feet long, and you want to plant one end solid in the ground. The other end goes up over the spot into which you plan to drill. You'll need another section of tree with a notch in it to act as an upright brace under your long pole. The free end needs to extend a couple of feet beyond your wellhead, for it's to that end, friend, that you'll attach a length of rope with a loop in the free end of it. Your tool string will be attached to the spring pole so that they move up and down as you put your foot in the loop and bounce up and down to drill your well. True, you'll have to readjust the tool string every couple of inches, but it's the cheapest way to drill a well and you certainly get a good workout while doing so.

Oil strikes occurred up and down Oil Creek and the Allegheny River. There is a story about James Evens, a blacksmith from Franklin. His house stood near a spring from which a viscid scum had often been wrung. During the dry season one year, he dug a seventeen-foot well, but the water smelled and tasted like petroleum. Well, sir, news of Drake's success set him to thinking. His thoughts flowed along these lines: "Drake sunk down a well near a spring that flowed with petroleum, and if it worked for him it might just work for me." Now, while he had no extra money to speak of, the owner of the local hardware store agreed to stand him the iron he needed to make his tools. It took him a couple of days to hammer out his tools, and a few more days to set up a spring-pole rig with his son Henry. Once they started

A driller placing a pike to mark the location of a new well.

drilling, things went well until they hit the seventy-two-foot mark. At that depth, their tool string slipped into a crevice and a piece of their drill bit broke off and got logged down at the bottom of the hole. After having no luck trying to fish out the broken piece, they decided to slap a hand pump on it and see if they'd struck anything good.

Well, sir, they hooked up the pump and began to pump vigorously. It took awhile, but soon they had a stream of fluid coming out that was viscous and dark green in color. It was heavy oil, smooth as silk, free from grit and coming out at a rate of twenty-five barrels a day. Now, if you thought the rhubarb that Drake's Well generated in Titusville was something, this was even greater. For you see, Franklin was the seat of Venango County and a city of some size already. The local citizenry all sped to see the well and the oil coming out of it. A doctor rushed over, forgetting not only his hat, but his coat and shoes as well. Ladies showed up without fixing back their hair or getting into their public clothes. At a nearby prayer meeting, a ruling elder who was taken with the events of the day was heard to send an imploration heavenward, calling on God to send down a shower of blessings and to make it a twenty-five-barrel-a-day shower. Truth be told, it was a red-letter afternoon and twenty-five barrels a day of thirty-dollar oil gave no one anything to sneeze at.

It was later that night when the long arm of the law showed up dressed in his best suit of clothes. One of the daughters of James Evens answered the door with the words, "Dad's struck ile!" Well, sir, the phrase quickly caught on and became a bigger hit than the well had been. It spread so fast and so far and wide that it became a part of the petroleum vernacular. In fact, it got to be so big that a song was written using that phrase as the title in 1865.

However, the tale doesn't end here. For you see, the young lady married Miles Smith, a furniture dealer from Franklin. Smith was originally from England and returned there for a visit in 1875. While at a party at a friend's house, the topic of discussion turned to Pennsylvania, and Smith was asked if he had ever heard the phrase, "Dad's struck ile!" The person asking the question had read it in the newspapers and never really believed that anyone had said it, figuring instead that it was the invention of some newsman somewhere. The assembled company was delighted to learn that not only had Smith heard the phrase, but also he'd married the girl who first said it. They say he was the toast of the party circuit for the rest of his stay in England.

Friend, that is but one of the many tales that occurred along the banks of Oil Creek and the Allegheny River in the decade right after Drake made his strike and launched an industry that changed the world.

THE LIFE OF A DRILLER

Now, all of this growth and industry would not have been possible without the people who lived and worked in the early oil industry. Friend, your life might not have unfolded like this, but the lives of many a man did.

When a fellar started out in the producing end of things, he began as a tool dresser, or "toolie." After he had learned the trade, he could then work as a driller or rope choker. This was the same no matter where you went. During the boom days, drilling went on straight around the clock. This meant that each well required a four-man crew, two rope chokers and two toolies. They worked in twelve-hour shifts that were called tours. These tours ran from twelve to twelve.

Now a toolie was there to learn the trade and help the rope choker. The toolie was required to keep up the steam pressure, climb the derrick to oil the crown pulley and generally make hissilf useful around the rig. But his chief duty above all else was dressing the bit. This meant that he was in charge of keeping the drilling bit shaped and sharp. He did this by heating it up in a hand forge and pounding it with a sledgehammer. You see, it was important that that bit had just the right shape or it might get stuck down in the well.

On the other hand, the rope choker had a highly important position. You see, in the absence of the producer or contractor, he was in full charge of the drilling. Getting the well drilled hung on his neck. The rope choker had to know how to keep the hole straight. This is not as easy a task as it might sound. First, you had to drill a round hole with a flat square bit. Not only that, but oft times the rock they were drilling through wasn't lined up nice and flat, but ran off at all angles. So the rope choker would sit on this high stool under the walking beam, holding onto this long, round stick that was lashed to the cable, and every couple of stokes he would give the string a turn. Now and then, he'd pull the string and send down the reamer to shape

Gib Morgan, an oil well driller of the first order and the minstrel of the oil fields

up the sides of the well. On top of everything else, he had to know what was going on down at the bottom of the hole, and all he had to tell him was the feeling he got through the cable. By and far, he was the most skilled worker on the oil field. Fact is, friend, there were stupid lawyers, stupid doctors, stupid businessmen, stupid professors and stupid people in general, but never was there a stupid rope choker. It is possible one might have existed somewhere, but he was a rare bird if he ever did exist at all.

Now, let us turn to the rig that they used and how it was set up. The derrick was about forty or so feet high and ten feet across on each side. Up top, on what was called the crown block, were placed two grooved pulleys, one for the sand line and the other for the drill cable. A short piece away from the derrick was the power plant. This was normally a six- to twelve-horse steam engine, with a belt that led to a wooden band wheel that turned a crankshaft. The Samson post was between the band wheel and the derrick. The Samson post is a huge upright timber with a walking beam swung to the top of it. One end of this walking beam was connected to the crank by a pitman. The other end was connected to the drilling cable and it was the motion of the walking beam that raised and lowered the string of tools in the well. Just inside the derrick and opposite the walking beam was a windlass. This windlass was made of a large wooden shaft and two wooden wheels. They were called the bull wheels and the bull shaft. It was around the bull shaft that the excess cable was wound after it had been spiked to a yard on the bull wheel. The free end of the cable ran up over the crown pulley and ended in a rope bucket to which was fastened the tool string. The cable was then fastened to the walking beam by a temper screw. Thus, by turning the screw, a rope choker could lower the string as the well went deeper. It was said that Uncle Billy Smith came up with the idea for the temper screw. It also proved to be a way of measuring how far you had drilled in a tour. For you see, each time you turned the screw a bit more of the cable came off of it, and when there was no more cable on the screw, you had to stop and reset the screw. Usually it was about a yard of cable per screw.

Now, as you might have guessed, wood breaks; eventually everything breaks. One of the first things that became apparent was that when a walking beam broke, the rope choker was in for a major headache. So in order to protect themselves, rope chokers and toolies alike started putting a big, strong timber across the derrick just under the swing of the walking beam. They called this beam a headache post, mostly because it prevented headaches when the walking beam broke. It also provided a gathering place for everyone to exchange news and information at the drilling site.

The long string of tools that were used to drill a well was normally just called the tools or the string. The tools included a chisel bit, a drill stem or auger, a pair of jars, a sinker and a rope socket. The string was there for the sole purpose of driving the bit farther down the hole. The bit was the business end of the string and was where the actual drilling took place. Now,

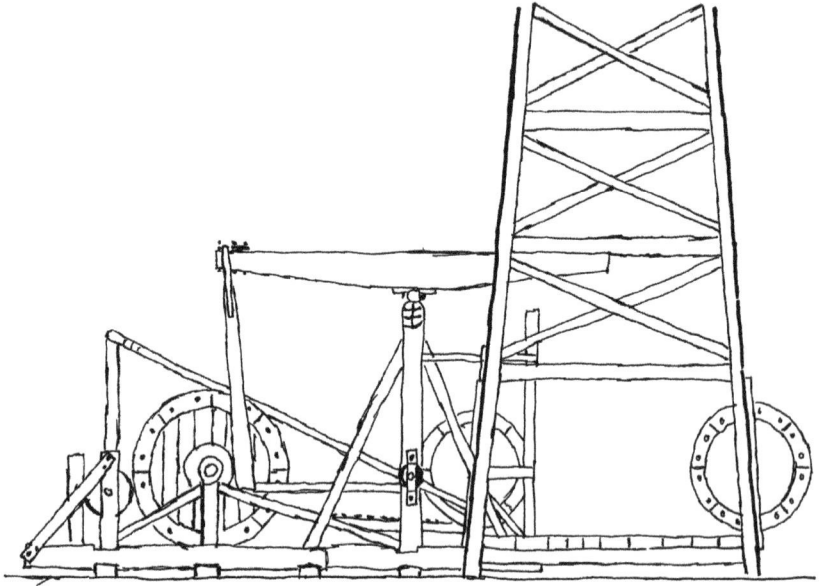

The base of a typical drilling platform.

early on, drillers started driving iron pipe down into the ground to prevent caving. The procedure of driving that casing into the ground was called "spudding down the well" and it was invented by Col. Drake. It was the only thing of value he ever gave to the petroleum industry.

Now, from time to time, you had to pull the string and clean the cuttings out of the well. Since they were drilling down through the earth, naturally they hit water well before they hit oil, and this water mixed with the cutting making a slush. After they had pulled the string, a long piece of pipe with a plug valve on the bottom end of it was lowered down into the well. This piece of pipe was used to bail the mixture of slush and water out of the bottom of the well. The process of doing this was called "bailing out a well," and the pipe was either called a slush bucket or a bailer.

As things will happen, the cable snapped from time to time and they would lose the string down the well. Oilmen had designed specific tools for getting the string back, too. There were rope spears for catching the cable itself, horn sockets for recovering a bit and long-stroke fishing jars if the bit was really stuck. It was rare that they just left the string down the well and walked away from it.

Now, friend, after the well was drilled, the final casing had to be set. This included shutting off the ground water by attaching a seed bag to the center tubing. A seed bag was nothing more than a round leather bag filled with flaxseed. When the seed got wet, it expanded out and sealed the space between the tubing and the sides of the well.

Well, sir, now you have a general idea of how they went about drilling a well and some of the tools and equipment they used to do it. The only other thing that you might be interested in is the way a rope choker lived and dressed.

You see, friend, the rope choker knew his own importance and behaved as such. If he was a drinking man and liked his whiskey, it was always good whiskey that he drank. On top of that, he was very particular about his dress. Normally he would wear a denim shirt, a pair of denim trousers, Wisconsin boots and a good hat. However, his hat must identify him as a driller. This means that even off duty, a rope choker's hat must have slush splashings on it. There are stories of rope chokers who took their new hats out to the slush pit and splashed them with sand before they wore them.

Friend, you might read somewhere that a rope choker would throw away the set of clothes he wore while he was drilling a well when it was brought in. While this did happen, don't think that the rope choker was being vain, as oft times if a well was brought in ahead of schedule or it was a good well, the producer would give the driller and the toolie a new set of clothes. Additionally, there was always someone waiting for that old set of clothes that the rope choker and the toolie were done with, even if Wisconsin boots did run a man twelve dollars a pair.

BOOMTOWNS

FROM EMPTY FIELD TO THRIVING CITY OVERNIGHT

One of the unique events that always seems to happen during a rush, no matter what the rush is for, is the emergence of boomtowns. To be sure, the Oil Region had its share of boomtowns. No sooner was a new strike reported then it would seem a new town or city had sprung up around it.

Up until 1859, there were some forty-three farms of various shapes and sizes along Oil Creek between the cities of Titusville and Oil City. The farmers made their way in the world by farming during the summer and hunting, fishing and lumbering during the winter. They were content with their lives, and if you had told them what was to come, they would have laughed in your face and gone about their business, wondering just what kind of a fool you were.

Yet, things did change, and they changed quickly, though not always for the better. What was once a forested valley soon became a valley of oil derricks and mud. In fact, the valley was often described as, "Wholly Unclassable, Almost Impassable, and Scarcely Jackassable"; the latter referring to the stories and tales about men riding down the muddy roads on horses, mules and jackasses, only to have the animals sucked out from under them by the muck and the mire, never to be seen again this side of kingdom come. Just how true the tales were, friend, I'll leave that for you to decide.

A good example of a boomtown is Petroleum Center. If you were to go there now, you would find a state park with a few buildings scattered here and there, some of them modern and some not. Additionally, there is a wooden sidewalk along a couple of the old streets with signs where buildings once stood and the start of a bicycle trail that follows the course of an old rail bed.

A wagon stuck in the muck and mire on its way to a new boomtown.

It was a far different story in 1864 when G.W. McClintock sold his two-hundred-acre farm to the Central Petroleum Company of New York. This company had been organized by Frederic Prentice and George Bissell. Within the McClintock Farm was a circular ravine that was about three quarters of a mile long. It was in this ravine where the foundations for Petroleum Center and Wild-Cat Hollow were to be found.

Well, sir, the Central Petroleum Company granted lease only to those whom it deemed to be actual operators, and then only for half the production in royalty and a large bonus. Brown, Catlin & Co. were granted one of those leases, and they drilled the well on the McClintock Farm. That well came in during August 1861 and was one of a multitude of wells with splendid records. Altogether, the Central Petroleum Company and its leases harvested over $5 million off of the McClintock Farm.

Is it any wonder, then, that people would want to start living that close to a gold mine? It was to that end that the Central Petroleum Company laid out half a dozen streets and started leasing building lots. Even though the lease on the lots was exorbitant, there were soon boardinghouses, offices, hotels, mercantile shops and, lest we forget, saloons and thirst parlors.

Nobody seemed to care that discomfort seemed to be the rule of the day. Hordes of people streamed into that forest of oil derricks. The food they found there was lousy, the beds were worse and let us not even begin to discuss the state of the liquors. It is enough to say that you could get drunk off of it; anything beyond that would be nothing more than sheer charity. It was stacking up to be a nice town. Unfortunately, the owners of the Central Petroleum Company were steadfastly against borough organization, and as such, the town traveled headlong into the future with a "go as you please, devil may care" attitude.

It was because of this attitude that prostitutes and sharpers were able to ply their trades without fear of either human or divine law. The dance halls were notorious and were often the scenes of violent outbreaks—on the good nights, that is. Around Petroleum Center, residents often told the tale of two gaming gentlemen who tossed dice with each other for twelve straight hours at $1,000 a throw.

In fact, The Center, as the locals took to calling it, became known as the "Wickedest Town in all of the Oil Country." It was known far and wide for its vice and lax view of things. Friend, you will understand this better when I tell you about one of the local madams. This particular gal worked her way up through the ranks until she owned her own "Free and Easy" and was worth over $150,000. "Free and easy" was but one of the names given to those houses where a man could go to let off his surplus wickedness and enjoy an evening in the company of a soiled dove or lady of the evening..

Things went well for The Center, but the handwriting was on the wall, and after only two or three years, the town was nothing but a shadow of its former self. By the turn of the century, it was all but gone.

The Center was one of many boomtowns that grew up around a strike, stayed active so long as the oil field stayed active and then died as the boomers, those men who follow the oil strikes, moved on to the next big strike. One of the unique features of all of these boomtowns was their names. While oilmen were not often creative in the day-to-day aspects of their lives, they sure did know how to name a town.

Why, even today, if you were to take a trip along some of the byways of the Oil Region, you can still find remnants and signs for such towns as Cash-Up, Funktown, Shamburg, Pioneer, Allemagoozelum City and Pithole, to name but a few. The buildings in these towns were all built pretty much the same way. On the whole, most of these buildings were called slap-ups.

Now, to build a slap-up is fairly simple. After securing your building lot, you must decide how big you want your building to be. Once you've figured

General Jesse L. Reno, the man who became the highest ranking officer from Venango County during the American Civil War and for whom the town of Reno, Pennsylvania, was named.

this out, drive a four-by-four post into the ground at each corner. Then, run a two-by-four rail around the top of the posts and another around the bottom of the posts. Now, figure out where you want your door and your windows, if any. On either side of these, run two-by-fours from the top rail to the bottom rail. The sides of the building are covered with one-by-six-inch rough-cut planks nailed up vertically. To finish off, put on a sloping roof, hang your doors and put in your windows, if you chose to have them. Whether you paint the building is up to you, but a coat of whitewash will help the wood last longer. As for a floor, well, that's also up to you, but it should take you no more than a couple of days to build a slap-up. When you're ready to move on to the next boomtown, you can either leave it, unload it on some other driller or tear it down and take it with you. A lot of people did just that, and buildings moved from town to town and strike to strike for years.

It was often surprising to newcomers in the Oil Region just how quickly following a new strike a boomtown could grow up and just how fast it could vanish after the strike paid out.

THE FRAZER WELL AND PITHOLE

Pithole Creek flows into the Allegheny River eight miles north of Oil City, Pennsylvania. It was, therefore, a likely choice to expand the search for petroleum to that area after all the land on Oil Creek had been leased and was being worked.

It was in 1840 that Reverend Walter Holmden purchased two hundred acres near the headwaters of the creek and moved his family there to farm. His son, Thomas, leased part of the property to the United States Oil Company for twenty years beginning in 1864. I.N. Frazer had started the United States Oil Company with the money he made as part owner of the Reed Well on Cherry Run.

In the late fall of 1864, Frazer set up a crude derrick in the woods below the Holmden House and began drilling. Workers drilled through the winter months, and in January 1865, at a depth of six hundred feet, they hit the sixth sand. At that point, the drillers started to watch the drill closely, as the felt they were close to an oil-bearing stratum. On January 7, at a depth of ten feet more, a torrent broke loose and the petroleum shot up into the air. This was the first oil gusher of the industry. Well, sir, that well flowed at a rate of 650 barrels a day, every day, until November 10, when it dried up.

As Frazer was drilling his well, two fellars started a company, Kilgore & Keenan, which they named after themselves and were drilling two wells not far away. These wells came in on January 17 and 19 at a rate of eight hundred barrels a day. At that point, the boom was on and it was only a matter of a few months before the city of Pithole stood where there had been naught but forest and farmland the year before.

It happened like this: a Mr. Duncan and a Mr. Prather got together, bought the farm from Thomas Holmden for $25,000 and began to lay out a city, dividing the land into half-acre leases that started out at $850 a year. When May arrived, all that existed in Pithole were three producing oil wells,

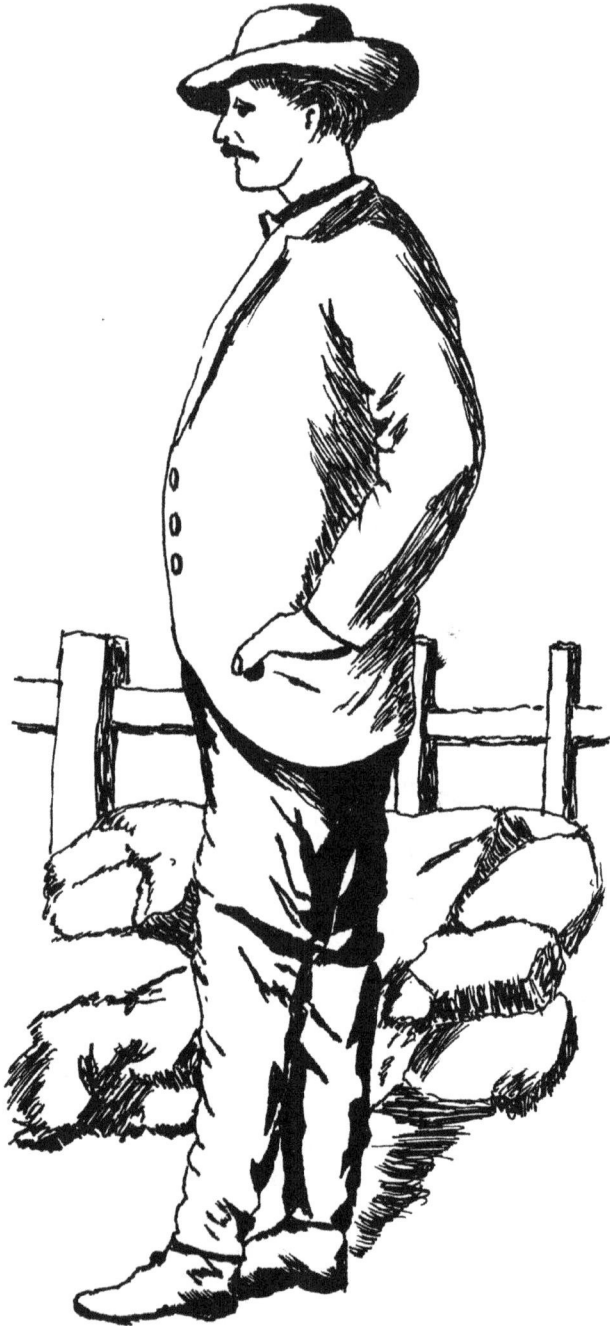

John McKeown, one of the thousands of men who came to Pithole to make his fortune and then made it many times over.

one well being drilled and three homes. That was not to last, for ninety days later Pithole boasted sixteen thousand residents and more conveniences and luxuries than many an older established city from back East.

Well, friend, those people were all there to make money in oil, and the speculations were flying fast and furious. The value of the lands around Pithole jumped and jumped and jumped again as speculators bought and traded leases and drilling rights with one another. There is even a tale of one operator who was able to sell seventeen-sixteenths of the shares in one well alone. A half-acre oil lease on the Holmden Farm often realized a bonus of $24,000 before a well was even drilled on the property.

Part of the reason for the extraordinary growth of Pithole was the end of the American Civil War and the flood of men and money that it brought into the region. The post office that was set up in Pithole quickly became the third largest in the state, behind only Pittsburgh and Philadelphia.

It was soon easy enough for one to find a good time in Pithole, for the town had hotels, theaters, saloons, gambling rooms and free and easies by the score. The more civil-minded folks organized a fire department and elected a mayor. To record all this, a daily newspaper was started up.

The newspaper recorded the fact that the best hotel in town was the Chase House and it was furnished for $100,000. That newspaper also carried the names of the nationally famous entertainers who performed at the eleven-hundred-seat Murphy Theater on First Street.

All the while, oil wells kept coming in. In June 1865, The Homestead Well came in gushing. The Deshler Well started flowing in August at 100 barrels a day, and The Grant Well came in right behind it at 450 barrels a day. These wells represent the tip of the iceberg, as each day saw another new well come in. In order to get the oil to the refineries, over fifteen hundred teamster teams were employed. They transported the oil right up until Van Syckel finished his pipeline from Pithole to Miller Farm and put them all out of work.

The teamsters were rightly upset at this turn of events because they had been getting three dollars a barrel to haul the oil from Pithole to Miller Farm or Titusville, and their wagons could hold between three to five barrels. Not only that, but it was often possible for them to make more than one trip in the course of a day. Now, friend, you would have to agree that it is hard to give up this kind of money when the average workingman was only making three dollars a day.

Still, the sixty hotels in Pithole could not accommodate the influx of people that summer. The rule became first come, first served, and many wound up sleeping on a plank or a pile of hay in the livery stable.

The Chase Hotel in Pithole was built and furnished in 1865 for $100,000.

It was during the summer of 1865 when Ben Hogan first made his appearance in Pithole. Even today, he is a man of some notoriety in the Oil Region and entire books could be, and have been, written about him. When he showed up in Pithole, he first made his way by taking on all comers under the Marquis of Queensberry rules. He was a fighter and a scraper from the word "go." Soon, he had pulled together enough money to open a free and easy of his own. To do so, he teamed up with French Kate, the self-proclaimed widow of a slain Confederate officer. They set things up so that Ben would take care of one's vertical needs and Kate and her girls would take care of one's more horizontal needs. It is said that from the time their Free and Easy opened until Ben and Kate closed the door in Pithole and moved on to the next boomtown, the receipts were never less than $1,000 a day.

Now, I don't want you to get the idea that there was no moral backbone in Pithole, for there was and it was a strong one. You see, friend, along with all the bars, taverns and dens of vice, just as many churches and schools were set up to educate the masses. Not only that, but there were several civil and fraternal organizations that had chapters in Pithole; a couple even began there. One of these was the Swordsman's Club. This was a social club that was set up by several of the leading, well-respected oilmen of the day.

The Swordsman's Club hosted many promenade concerts and elaborate balls. The membership was made up of the gentlemen of the Oil Region—the producers, financiers and gentleman speculators who had made their fortune. It was such an elite group that invitations to the events it hosted were always sought after by those who wanted to show that they had "made it" in oil. While the club only lasted for two years, from 1866 to 1868, it counted among its members four congressmen and two ex-governors, as well as numerous notable men who went on to make names for themselves outside of the Oil Region.

Now, don't go thinking that it was only the people in the upper levels of society who set up these organizations. The Forty Thieves, itself a famous social and charitable organization, was set up by the well superintendents, producers and drillers. The name came from the way that distant stockholders, who really didn't understand the petroleum industry, referred to these people and blamed them when there was a lack of dividends on their stock. As much of the membership was made up of young men who took a great delight in the name, the group was known for its practical pranks and habit of helping out those in need.

It was in the fall of 1865 when production on the wells around Pithole started dropping off. As I said before, The Frazer Well stopped flowing altogether on November 10. One by one, the other wells started drying up and shutting down.

Though by March 29, 1866, a person could buy a through pass and travel by rail to Pithole from anywhere in the country, the city was not to last. There had been a fire in February that had wiped out the Tremont Hotel and several of the adjoining buildings. Then, in May, fire came back again for a second attempt on Pithole. This time it was more successful, burning down some eighty buildings and spreading to over thirty wells and some twenty thousand barrels of oil.

At that point, Pithole's days were numbered. The best buildings were torn down to be rebuilt elsewhere, and one of the jewels of the city, the Danforth Hotel, was sold for sixteen dollars as firewood. While there were still six registered voters living in Pithole in 1876, most would agree that the true life of the city, from birth to death, lasted a mere five hundred days.

THE TRANSPORTATION OF PETROLEUM

With all that petroleum being pumped out of the ground, a method had to be found to move it from the wellhead to the refinery and then on to the final consumer. Oilmen in the field didn't have much interest in the second part of the equation, but they were certainly concerned with the first part.

In the very beginning, way back before Drake's strike outside of Titusville, what petroleum there was could be moved in jars, bottles and small wooden casks. It would be an even bet to guess that Kier bought his petroleum in casks or small barrels from the wells around Tarentum, Pennsylvania. After he had it in his possession, he would either pour it into half-pint bottles to sell as Kier's Medicinal Petroleum or refine it into Carbon Oil and package that in barrels to sell along with his lamps.

It was when Uncle Billy Smith made the strike for Drake that a better method had to be found. Drake's Well was producing twenty barrels a day, and oilmen figured a barrel at 40 gallons plus 2 to make up for loss during transportation. Thus, twenty barrels would work out to be 840 gallons of petroleum each and every day. Is it any wonder that the men who made barrels, coopers by trade, were in great demand and made a good living in the Oil Region?

It was not a long time after things got rolling that the oilmen started to build huge wooden stock tanks at the wellhead, or very nearby, to hold the petroleum in bulk until they could barrel it up and move it to the refinery. This solved one problem, but not the other. They still had to get their petroleum to the refinery. This could either be done in a single move or in a series of moves in which they first moved the barrels to a railhead or flatboat

A couple of flatboats on Oil Creek awaiting their next load of petroleum to take it downstream to the refinery.

dock and then from there to the refinery. It all depended on where their well was and what was available to them.

The first way that the oilmen figured to move barrels of petroleum was in the wagons owned by the farmers whose land they had leased to drill their wells. The job normally fell to the older boys in the farmer's family and they were paid an average of three dollars for every barrel of petroleum they delivered to the refinery. The wagons they were using were the pickup trucks of the day. They were flat bottomed, with sideboards that came up to hold the cargo in place. The back was closed off with either another board, a rope or a net of some sort. There was a seat in the front raised above the level of the wagon bed, and the whole rig was pulled by one or two horses, depending on the size of the wagon. Well, friend, if you figure that those wagons could hold from three to five barrels of petroleum on average for each trip, you can begin to see how much these boys liked the additional work.

Another method used during the early days of the industry was to take a flatboat and caulk the seams between the boards so that it was nice and tight. A standing board was placed across the gunwales because the entire boat was filled with petroleum. It was not in barrels or any other form of container beyond the boat. The boatman stood on the board and guided this floating pool of petroleum downstream to the refinery. I have no doubt, friend, that you can see all sorts of problems with this method of transportation. However, what you don't know, unless you happen to be familiar with the Oil Region, is that Oil Creek is only high enough to accomplish this trip during the spring thaw. That fact alone made things a lot dicier, as the higher waters made it more likely that the boat would tip and spill all that oil into the creek.

Well, fear not friend, they started using barrels to keep the petroleum better contained while it was being transported down Oil Creek. The other problem of only being able to boat down the creek during the spring thaw had been solved years before by the lumbermen. What they did was to place dams on the streams that flowed into Oil Creek and allowed the water to pool behind them into ponds of a decent size. Then, on a published day at a given time, each dam was opened, flooding Oil Creek and raising the level of the water enough so that one could navigate down to Oil City and the Alleghany River. This artificial flood was called a freshet, and it proved to be a popular way of transporting petroleum.

Not that this system wasn't without its problems, the main one being that there was so much petroleum to transport that the number of boats on Oil Creek was such that you could often walk from one bank to the other without getting your feet wet. Of course, oilmen being what oilmen are, they couldn't do things in a civil and organized manner. No, sir, they were always jostling each other and bouncing off of each other, trying to make sure that theirs was the first boat to get to the refinery or the docks in Oil City during each freshet.

From Oil City, the petroleum that was heading onto Pittsburgh and other points along the way was loaded onto steamships and carried in that way. The first petroleum to be shipped this way was aboard the *Venango*, captained by Bill Phillips. He went on to greater fame when he became a producer, and his well, the Phillips No. 2, came in at four thousand barrels a day.

Getting back to the freshets, they often raised the level of Oil Creek anywhere from twenty to thirty inches and there were often as many as two hundred flatboats carrying up to thirty thousand barrels of petroleum. Of this, only about a third ever made it to the refinery. There are those who claim that a third of it was lost by leakage and the other third was lost through accident. Not having been there myself, I can't say for certain if this is so, but that is what is recorded in the sources of the day.

Dunsmore oil tank railroad cars were built by placing wooden stock tanks on flatbed rail cars. The idea was so successful that soon tank cars were designed that more closely resemble those we see today.

Well, friend, taking all this into account, and given the fact that most oilmen are born tinkers, it was no wonder that better ideas were soon developed. It was Amos Dunsmore who came up with the idea of putting two stock tanks on a flat rail car and then transporting the petroleum inside those stock tanks. His first cars were built in 1865 and they were such a success that railroad cars designed specifically for transporting petroleum in enclosed metal tanks were on the tracks by 1869. This method of transporting petroleum has improved in safety over the years and remains one of the two primary ways we transport petroleum across the country today.

The other way began as an idea late in 1860. General Karns of Parkersburg, West Virginia, suggested to C.L. Weaver of Bradford, Pennsylvania, the idea of having petroleum flow down the side of a mountain inside a six-inch pipe. Unfortunately, the American Civil War stopped them from putting this idea into practice. A year later, the idea resurfaced to lay a wooden pipe in a trench from Tarr Farm to Oil City. This time a bill was introduced in the state legislature to provide a general pipeline charter and authorize the construction. And this time the teamsters, four thousand of them, put a halt to it.

The next person to step up to the plate was J.L. Hutchings, who brought a rotary pump, on which he held the patent, to the Oil Region, and in 1862 he laid a pipe from Tarr Farm to the Humboldt Refinery below Plummer. With it he demonstrated that the idea would work. However, the joints of his pipeline leaked like a sieve and the teamsters once again stepped in to destroy the works so that their livelihood would not be threatened.

It was in 1864 that the transportation problem was finally resolved. That year, Samuel Van Syckle came to the Oil Region from New Jersey. In August of that year, he completed a two-inch line from Pithole to Miller Farm, a distance of some five miles. It was with the completion and acceptance of this pipeline that the oilmen had finally solved the problem of safely transporting their petroleum.

THE POND FRESHET DISASTER

During most of the year you can walk across Oil Creek wearing nothing more than calf-high boots and not get the least bit wet. Of course, there are times during the year when Oil Creek runs fast and furious and can be navigated both upstream and down.

The farmers, who also made money by timbering their lands, had devised a system of dams on the streams that feed into Oil Creek. It was by backing up the water behind these dams that they could decide when to flood Oil Creek and, as such, have enough water depth to float their timbered logs down to the sawmills. It was this system that the oilmen adopted for their own purposes. Instead of transporting lumber, they would use the artificial flood to transport petroleum down to Oil City, where it could be loaded onto steamboats for the trip to Pittsburgh.

These artificial floods were called freshets, and they were often overseen by Reverend A.L. Dubbs. It was he who set the day and time for the gates to be opened. When the gates were opened, the water rushed forth and raised the level of the creek from two to three feet. The boatmen, who had been standing ready, then cast off their lines when the current was just right. It required sound judgment and perfect timing, for if a loaded boat cut loose too soon, it ran the risk of grounding and being trapped under the oncoming rush of boats. On the other side of things, if the boatman cut loose too late, then the boat ran the risk of being left behind until the next freshet.

The amazing thing, friend, was that the entire trip was only a distance of, at most, eleven miles. This was the distance from The Noble Well to Oil City. The Noble Well was as far up Oil Creek as the boatmen were willing to go.

Now, the boatmen were all skilled river pilots and earned $100 to $200 per trip. These men had the skill to avoid the snags and rocks that were to be found throughout the length of Oil Creek. There were basically

A small backup of flatboats during a pond freshet along Oil Creek.

two different kinds of boats that were used most often in this endeavor. The smallest were the Guipers. These were small, scow-shaped craft that were between 20 to 50 feet in length and 5 to 10 feet wide. They carried twenty-five to fifty barrels of petroleum in bulk form in their open holds. The other boats were the French Creekers, which were anywhere from 80 to 120 feet long and 15 to 20 feet wide. They were designed to carry ten to twelve hundred barrels of petroleum in bulk or barrel form. The other thing you need to remember, friend, is that both types of boats had shallow drafts of not more than twenty inches.

The use of barrels depended on the price of the petroleum. When the price was low, the petroleum was shipped in bulk. That is to say, it was poured into the open hull of the boat and floated down Oil Creek that way. However, the slightest change in the motion of the creek could cause the boat to capsize and dump the entire contents of the hull into the creek. When the price of petroleum was high enough, the producers used barrels to minimize their chances of loss.

The freshets usually happened twice a week, usually on Wednesdays and Saturdays. They were not cheap, for each freshet cost the suppliers between

A typical well pumping oil into a stock tank. When the tank was full, the oil was transported to Oil Creek to be shipped down to Oil City with the next freshet.

$200 and $300. This money was paid to the mill owners for storing the water and the use of their dams.

With a cry of "Pond Freshet!" the dams were opened. As the cry echoed down Oil Creek, those who could came down to the banks of the creek to watch the fun. For you see, not only were property and life imperiled, but boats were often ground to fragments. When that happened, thousands of barrels of petroleum would spill into Oil Creek to be lost upon the water. That floating petroleum was up for grabs—it could be gathered up by anyone. Therefore, people were always on hand to begin their fortunes by collecting the spilled petroleum. Not only that, but due to the variety of backgrounds of the people of the creek, one was often likely to learn a new word or phrase that was not quite proper Sunday go-to-meeting talk.

The reason for all of this was that if a boatman tried to get a jump on the freshet, he would find himself running ahead of the rising water. If this was the case, he might find himself running aground on the first shallow place he came to. Once a boat was aground, the current of the freshet often swung it around broadside to the flow of Oil Creek. When this happened, the boat soon filled up with water and became a snag for the other boats coming along behind it.

If it was a Guiper that had gotten stuck, it was soon to be a destroyed mess of lumber and petroleum flowing down Oil Creek. A big French Creeker would likely be the beginnings of a new temporary dam across Oil Creek.

If a boatman was coming downstream during a freshet and saw that there was a boat snagged up ahead, he was far better off being in a French Creeker than in a Guiper. The reason for this, friend, was that between how fast the current was flowing and the amount of other boats on Oil Creek, it was far more likely that he, his boat and his cargo would survive the collision if he had a larger boat. It was truly survival of the fittest, as the smaller boats would be run over and destroyed by the larger boats and the weaker would be crushed by the stronger.

Eventually, one of the French Creekers would find its way past the sunken boat and then all the boats that could would follow through the gap that had been made in the jam. Those that couldn't make it through or got hung up somewhere else by the lowering freshet waters had their petroleum removed and placed on the creek side while the boat was salvaged and repaired to be ready for the next freshet.

In addition to the shallow parts of Oil Creek, there were several narrow places where jams often occurred. The most notorious of these could be found at the old Forge Dam, Clapp Farm and the sandbar at the mouth of Oil Creek. Additionally, the piers of the bridges at McClintockville and Oil City were also easy places to get hung up. For, if a boatman misjudged the distance and missed the channel current, he would find himself hung up crosswise on the pier. The size of the jam depended on where the boatman and his boat were in the fleet coming down Oil Creek that day.

This is exactly what happened on May 31, 1864, when one of the big French Creekers in the front of the fleet misjudged the current under the bridge in Oil City and wound up across the pier. As it was in the front of the fleet, there were somewhere between 150 to 200 flatboats coming down Oil Creek behind it that day. While some of the first boats got by the jam without a problem, most of the fleet was not so lucky. That day has gone down in history as the Great Pond Freshet Disaster because of the fact that not only did most of the flatboats riding the freshet that day get hung up in the jam, but also because over thirty thousand barrels of petroleum were lost in Oil Creek.

PIPELINERS AND THE
PIPELINE WAR

When Samuel Van Syckle completed the first pipeline from Pithole to Miller Farm, a distance of some five miles, he not only changed the way petroleum was moved, but he also touched off what became known as the Pipeline War.

First, let me set the scene for you. Up until Van Syckle laid the first successful pipeline, the main way of transporting petroleum over land was in barrels hauled by teamsters. They charged an average of three dollars a barrel to haul them from the wellhead to the railhead or flatboat dock. Not only that, but they had been successful in blocking prior attempts to lay a pipeline through which petroleum could be run.

Van Syckle's pipeline ran from The US Well along a right of way through the woods to Miller Farm. The pipe itself was bought in fifteen-foot joints and was tested to hold nine hundred barrels to the square inch. Each joint cost Van Syckle and his partners fifty dollars. The joints were lap welded together, and while part of the pipeline was buried at a depth of up to two feet, part of it ran along the top of the ground. Once it was completed, three steam pumps were used to push the petroleum through at a rate of eighty-one barrels an hour.

As the completion of this pipeline declared an end to the reign of the teamsters, they looked on it as an act of war. Since they had been able to derail past attempts to set up a pipeline through sabotage and intimidation, they figured it would work the same way again. As such, they got ready to take the battle to Van Syckle.

The first time that the teamsters came up against this new technology was when Henry Harley began laying two separate pipelines in 1865, one

A teamster's wagon resting outside an oil well, waiting to be filled with barrels for transportation to Oil Creek or the railhead.

between the Tarr Farm and Plummer and the other between The Noble Well and Shaffer. Both of these pipelines were made out of two-inch pipe and could deliver up to two thousand barrels a day. This fact upset the teamsters to no end, and they decided that the pipeline could not be allowed to continue. Thus, they used their teams and heavy chains to pull the joints apart. Harley, however, was determined, and he relaid his pipes, finishing again in 1866.

This time, the teamsters threatened Harley's life, taking torch to his large wooden stock tanks in the middle of the night. This fire created a frightful scene that damaged not only the shipping platform, but also four tank cars and all the petroleum that was in the tanks. By this time, everyone had decided that it would be best to arm themselves. This was probably a good idea, for the teamsters returned on another dark night. This time, besides setting additional storage tanks on fire, they threatened the lives of the men Harley had hired to keep watch over things. This time, friend, shots were fired and one of the teamsters lost his life.

As events of this type often do, this one caused fear to spread throughout the Oil Region. The oilmen took extra care, and their families worried about them all the more. Friend, all of this was with good reason because there were soon fires in Titusville, Pithole, along Oil Creek and even as far away as Franklin. It would not have been as bad as it was except that when the oilmen were building their boomtowns, they built everything close together, often using one wall for two buildings. Because of this, when the oil-producing structures burned, homes often burned as well. There was plenty of damage and more than a few fatalities. It got so bad that a vigilance committee was formed in and around Titusville, and one of the first things it did was to erect a prominent gallows as a warning to all.

While the teamsters were keeping up their campaign against Harley, Van Syckle was able to get his pipeline laid from Pithole to Miller Farm. As soon as the teamsters became aware of what Van Syckle was up to, they turned their attention on him as well. It began with nothing more than putting up signs around Pithole and on the trees along the pipeline. These signs were meant to discredit the idea of pipelines and often condemned the management. When these didn't quite do the job, the teamsters began breaking up the pipeline with pickaxes and scattered the joints throughout the woods. This tactic didn't work too well either, as Van Syckle was determined and kept at his pipeline. It finally got to the point where Van Syckle ordered carbines from New York and hired men to patrol his pipeline with orders to shoot on sight anyone who attempted to damage his pipeline.

On October 10, 1865, Van Syckle completed his pipeline and, between his work and the work of Harley, the transportation of petroleum changed forever. True, the pipelines did put the teamsters out of work, but that is the nature of progress, and the teamsters either found other work in the Oil Region or moved on to other pastures.

NITROGLYCERINE AND THE ROBERTS TORPEDO COMPANY

In 1846, an English chemist figured out a new way to combine fuming nitric acid, sulfuric acid and glycerine. The compound he came up with was called nitroglycerine and initially was sold for homeopathic and cosmetic purposes. However, when a small batch that had been shipped to New York City as a specimen exploded, it caused quite a stir.

This event would have gone unnoticed in the Oil Region except for the work of Henry Dennis and William Reed. For you see, in 1860, Dennis drilled the first well at Tidioute, Pennsylvania. Unfortunately, he got his tools stuck at the bottom of the well and could not fish them out. To rectify the problem, he got himself a three-foot length of copper pipe two inches in diameter. Then, he plugged one end, filled it with gunpowder and inserted a long fuse cord in the other end. He lowered this pipe down into the well and exploded it in the presence of six men. The well filled up with water, oil and bits of rock. On top of that, the smell of petroleum in the air was so strong that people coming up the hollow noticed it long before they got to the wellhead.

That same year, John F. Harper was hired to explode five pounds of gunpowder at the bottom of A.W. Raymond's well near Franklin, Pennsylvania. Because the tin case he used on his first attempt collapsed before it got to the bottom of the well, he gave up on the idea. However, William Reed had assisted on the project, and he was able to work with the idea and develop the process further. What Reed came up with was eventually called the "Reed Torpedo," and it was first used on John C. Ford's well in 1866.

Reed put five pounds of powder in an earthenware bottle and attached it to a string of gas pipe. It was with the gas pipe string that the jar was lowered down into the well to a depth of 250 feet. The powder was set off by dropping a red-hot chunk of iron down through the inside of the gas pipe. The resulting explosion threw water out of the well and shot the gas pipe up with enough force that it not only knocked down the walking beam and the Sampson post, but also agitated the water in Oil Creek. As petroleum came out of the well with the water, tubing was put down the well and Ford began pumping the well again. This was the first successful torpedoing of a petroleum well.

As early as 1860, William F. Kingsbury advertised that he could blast petroleum wells in order to increase their production. There were several others who claimed that they could do the same, and between 1860 and 1865, there were somewhere around fifty wells that had gunpowder used in them to cause an explosion for the increase of petroleum production. However, there was a glut of petroleum on the market and it was fairly inexpensive to drill another well, so the torpedo industry languished.

It was with a complete disregard to all that had come before him that, in November 1864, Colonel E.A.L. Roberts filed a patent for "a process of increasing the productiveness of oil wells by causing an explosion of gunpowder or its equivalent at or near the oil bearing point, in connection with superincumbent fluid tamping." You see, friend, he claimed that it was the action of a shell at Fredericksburg in 1862 that had exploded in a millrace that had given him the idea for the process. Unfortunately, he could never show proof that he had been in Fredericksburg on the date he claimed and he furnished no papers or drawing for his application for over two years.

Regardless, he constructed six torpedoes and brought them to Titusville in January 1865. It was on January 21 that Captain Mills gave Roberts permission to test his process in The Ladies Well. Roberts exploded two torpedoes and the well flowed with fresh petroleum and paraffin. Still, he was not out of the woods yet, as there were several other patent applications on file and the whole matter was tied up in the courts for two years. In the end, everything was decided in Roberts's favor, and by such matters he was granted a monopoly of the whole process.

That having been settled, Roberts set up a company in New York to construct the torpedo shells and set about convincing producers that his technique would not damage their wells. It was when he shot The Woodin Well on the Blood Farm in December of 1866 and it began pumping eighty barrels a day that everyone started to come around to seeing things his way.

Col. E.A.L. Roberts, the inventor of the process of using a nitroglycerine torpedo to restart a well that has stopped producing, as well as the founder and owner of the Roberts Torpedo Company.

It was during the following year, 1867, when the demand for Roberts's services began to increase. The road was not smooth, as he was involved in many lawsuits based on people infringing on his patent. Since the justices seemed to be on his side, Roberts started to win the decisions even though a collaboration of producers had pooled together $50,000 to break his patent. It was around this time that Roberts substituted nitroglycerine for gunpowder. Seeing the results of this switch, he set up a factory in Titusville specifically for the manufacture of the explosive.

It wasn't long before the torpedo disputes became all-out war. The Roberts Torpedo Company charged a producer $200 just for a medium shot. If you needed a bigger one, then you could figure on paying more. It was because of this that the moonlighters came into being. These men would charge you a lot less than Roberts and would appear at your well in the middle of the night. Then, by the moonlight, they would shoot your well for you. Roberts was having none of this, and he hired a legion of spies to report on any producer who even thought about hiring one of these moonlighters.

Once Roberts knew for certain that a producer was trying to sidestep his hard-won patent, it was off to court and another lawsuit was filed. It is said that over two thousand of these prosecutions were threatened, mostly against producers who hadn't toed the line. Roberts had an army of lawyers working for him and they always made sure to have the trial location set in Pittsburgh. It was because of this, the enormous costs of attorneys and the high settlement amounts of those cases that did go to trial that most producers sought a settlement of some sort. So, friend, it was more by lawsuit than by the actual shooting of wells that Roberts and his brother became very rich.

In June 1873 Roberts's patent was reissued, causing an even more burdensome load on the producers of the region. It got to be so bad that the suit brought by Roberts against Peter Schreiber in November 1876 was the beginning of the end for the monopoly held by the Roberts Torpedo Company. For you see, friend, Schreiber had copied the unpatented torpedo design of the Cocker cartridge and Roberts wanted his head.

It took three years for the case to reach trial, and when it did, it was a glorious event to behold. Schreiber had hired three attorneys to defend him: Henry Baldwin, James C. Boyce and the Honorable J.H. Osmer, all of Oil City. It was during the time between the filing of the suit and the beginning of the trial that Mr. Boyce worked up a masterful defense. He was able to prove beyond a shadow of a doubt that explosives had been used in both water wells and petroleum wells for years before the government had granted Roberts his patent. Not only that, but the methods were so similar that one would be hard pressed to find any differences between them. Unfortunately, the presiding justices for the case, Justice Strong and Justice McKennan, had been hearing cases of this type for several years and had always ruled in favor of Roberts. Thus, they were not about to change their minds at this point, and again the trial went to Roberts.

Still, as they say, you can win the battle and lose the war, and this is exactly what happened to Roberts. All the research that had been presented during the trial had shed light onto the subject that had not shone before. This brought even more evidence to light, and the total mountain of historical data was enough to prevent Roberts's patent from being extended a second time. This opened the field to other companies and greatly reduced the amount it cost a producer to shoot his well in order to bring it back into production.

THE MOONLIGHTERS' WAR AND THE FATE OF NITRO SHOOTERS

When Col. E.A.L. Roberts secured the patent for exploding nitroglycerine in the bottom of a petroleum well with an eye toward improving the production of the well, he touched off one of the most dangerous conflicts that ever occurred in the petroleum region.

The conflict was called the Moonlighters' War, with the producers who couldn't or wouldn't pay the rates that Roberts demanded to shoot their wells on one side, and the Roberts Torpedo Company on the other. In the middle of this conflict were the nitro shooters themselves.

The nitro shooters were the men who did the actual work at the well. They either worked for Roberts or they didn't, and sometimes they did both. When they worked for Roberts, they did their work by the light of day and when they didn't, they did their work by the light of the moon.

No matter who they worked for, shooters traveled around the petroleum region in specially designed buckboard wagons. In the back of the wagon were the empty torpedoes, assorted pipes, cables and other tools that they might need at the wellhead. They carried the nitroglycerine in tin cans that were about five inches wide, five inches deep and twenty inches tall. These cans were put in felt-lined cells that were directly under the front seat of the buckboard. Everything would go smoothly as they traveled around, unless, of course they hit a bump and jarred the nitro.

This is what happened to Doc Haggerty in December 1888. He was a teamster employed to transport nitroglycerine to the magazine near Pleasantville, Pennsylvania. He was seen that day at the magazine and then twenty minutes later an explosion shook the valley. While they found pieces of his horse and his wagon, absolutely nothing was found of Doc

One of the men who worked for the Roberts Torpedo Company charging a torpedo with nitroglycerine prior to lowering it down the well.

Haggerty. He had vanished so completely that some say we'll see him again when kingdom comes. The sad part of this event was that the insurance company with which he had a $5,000 policy resisted paying on the policy on the grounds that, since no remains of the allegedly deceased man could be found, he might still be alive somewhere.

When the shooter finally made it to the wellhead, the hard work really began. At that point, he had to make sure that the drill string and all the tubing had been pulled out of the well. Once he had ascertained this fact, he began his work. The first thing he had to do was hook up the torpedo to a cable so that it could be lowered to the bottom of the well. When the torpedo was connected, he then brought the first can of nitro from his wagon and very carefully poured the nitro into the torpedo. He had to be very careful because the slightest jar could cause the nitro to explode. Once the torpedo was full, it was lowered to the bottom of the well. Depending on the charge ordered, that might be the only torpedo put down the well or it might not be. There are accounts of as much as one hundred quarts of nitro being used in a single well.

Now, friend, when the well is ready to shoot, all that has to be done is to explode the nitro at the bottom of the well. In order to do this in the most efficient way possible, shooters in the Butler Field came up with what was called a go-devil in 1876. This was a weight that was either dropped to free-fall down a well, or a ring placed around the taunt cable to be dropped that way down the well. Whichever way it was dropped, the same thing had to happen. As soon as you dropped that weight, friend, you had better go like the devil himself was after you. It was this act that gave the name to the weight. The reason for running away like a scared schoolgirl was that seconds after the nitro exploded, the air above the well would be filled with a shower of water, oil, mud, slush and stone chips. This would saturate the derrick on the way up and soak everything for a great distance around the well when it came back to the ground.

Torpedoing wells was a dangerous business, as the tale of Doc Haggerty should have made clear. A professional shooter had to have nerves of iron, partake lightly of the drink and be keenly aware of the fact that the slightest misstep could send him to his next life. One of the best was a shooter for Roberts, James Sanders, who shot six thousand torpedoes in his career without a single accident. On the other hand, William Munson achieved the notoriety of being the first victim of nitroglycerine in the summer of 1867. He entered his workshop and storehouse below Reno one morning in August. After that, no one is really sure what happened until the explosion was felt all the way up in Oil City, a distance of several miles. While there was nothing left of his buildings, bits and pieces of Munson covered several acres of ground. After the good folks of Reno gathered up the little pieces, they placed them in a small box and sent it to his family in New York for interment.

One of the best examples of why shooters were advised to stay away from the drink occurred in 1879 near the town of Red Rock. A moonlighter by the name of Reed came in to town one night more than a little in his cups and carrying two ten-quart cans of nitro in a meal sack slung over his shoulder. Moonlighters carried their nitro this way in order to keep their actions hidden from the Roberts Torpedo Company. Well, friend, the people who were there watching the events of the evening unfold were more than a little concerned to see Reed so far along the way to being drunk. There was not much they could do, as it would have been even more dangerous to try and wrest the nitro away from him. I'm sure they breathed a sigh of relief as Reed headed up the hill toward the wellhead. Still, those people surely hung around expecting an explosion at any time and they were not disappointed. The explosion shook the countryside and left a hole where Reed had stumbled and fallen under a tree that was ten feet in diameter. What was left of Reed was found mostly in the tree, but also parts of him could be found over an area that was almost an acre in size. While Reed was not that good of a shooter, the tale of his death was often told as a way to show new shooters what could happen if they mixed nitro and whiskey.

Nitroglycerine was also involved in one of the strangest tales in the entire Oil Region, which occurred in Rouseville. It was late in December 1869 and an unknown shooter had just taken a two-quart can of nitro out of the magazine and headed up the hill. Before he got to the well he was shooting that day, he had to stop at another well and make sure that everything was all right. Being a cautious man, he set the container of nitro down on the ground outside of the engine house as he made his way in to check things over. While he was inside, a rampant hog happened by and decided that the lardlike substance in the can would make a good dinner. Just as the shooter was coming out of the engine house, the hog was finishing his meal. After that everyone in Rouseville gave the greedy hog the widest latitude. In fact, it had the fullest passion of any sidewalk it chose to walk down. All the dogs in town were likewise confined for fear that they might chase the loaded porker against a post or wall and set it off. To the same end, no one was foolhardy enough to place a kick at the creature lest it should explode and send the town and all its belongings to their eternal reward. The matter became serious and the discussion of how to safely dispose of the important hog was the grand conundrum of the hour. Then, when the hog finally did go to his just rewards and was made into sausage and headcheese, a whole new terror was added to the long list that the Oil Region already possessed.

THE GREAT BENNINGHOFF ROBBERY

Prior to the oil boom, John Benninghoff was a farmer who owned two large blocks of land along Oil Creek just outside of Petroleum Center. Benninghoff had owned the land since 1836, long enough for the stream on one of the blocks of land to become known as Benninghoff Run. His was a family of farmers and, accordingly, they had the mentality of farmers.

When the rush started, he was more than willing to sign leases for a cash bonus and a liberal royalty on the product. He kept the leases to five acres each and refused to lease a certain piece of his land. Now, friend, he had many tempting offers to lease this piece of land, as it looked real promising to yield petroleum, but old man Benninghoff would not have his prized potato patch trodden upon. So it was that the rest of the leases were signed and the drilling began.

Friend, this was around 1866, and a series of wells being drilled on the Benninghoff Farm came in far above what was expected of them. In fact, The Lady Herman, which was drilled by Robert Herman and named after his wife, was not only a genuine beauty of a well, but it was also the first well ever cased and the first well attached to its own pumping station. Oil from that well was pumped up the hillside to the town of Shaffer, where it was sold to the refineries that were located there.

As I said, Benninghoff was a farmer one day, the next he was a rich farmer and on the third he was famous or infamous depending on how you look at things. You see, Benninghoff was a farmer and he had a farmer's distrust of banks, but he had lots of money coming in and he decided to store it at home. To that end, he went out and bought a cheap safe to hold it in. Friends warned him to be careful or thieves might break in and steal his money, Yet, he was still

An oil well of design and location similar to The Lady Herman Well.

looking at the world through the eyes of the farmer he had been before the oil boom. Therefore, he thought that the safe he bought was good enough and that, as they say, was that.

Now, talk of the situation spread as one might expect it would and it wasn't long before James Saeger of Saegertown, Pennsylvania, heard about the treasure being stored along Oil Creek in such a flimsy container. While Jim belonged to a respectable family and was himself a merchant in Meadville, the thought of all that cash so readily at hand was too much for him. To that end, he spoke with a friend of his, one George Miller, about how easy the safe would be to crack and then engaged the services of two burglars from Baltimore, fellars named McDonald and Elliot, to be precise.

In order to make sure that nothing would go wrong and the planned robbery would be a success, the three of them also enlisted the aid of Jacob Shoppert of Saegertown and Henry Geiger, who not only worked for Benninghoff, but slept in the house as well. Finally, to make doubly sure that nothing would go wrong, they planned to strike on a night when Joseph, Benninghoff's son, was to attend a dance.

Well, sir, on January 16, 1868, a Thursday evening, Saeger, Shoppert, McDonald and Elliot left Saegertown in a two-horse sleigh bound for Petroleum Center, a distance of some twenty-nine miles. They arrived around midnight and knocked on Benninghoff's door. It was Geiger who answered the rap and he was promptly bound and gagged as per their prior arrangement. The foursome then bound Benninghoff, his wife and daughter and proceeded to get to work on the safe. Friend, it was in very short order that they had that flimsy box opened for all the world to see. Now, at this point, they loaded all the cash into bundles, which were taken out to the sleigh, and then the devils had the nerve to not only sample Mrs. Benninghoff's pies, but to also drink a gallon of her milk.

It was about an hour or two after their leisurely departure that Joseph arrived home from the dance and freed his family. When they took a good look at what remained of the safe, it was found that the thieves had made off with $265,000, the bulk of which was in gold. Friend, we're talking about 1860s' money here—that figure has not been converted into modern dollar amounts. However, much to the relief of the Benninghoff family, it was further discovered that a package of large bills done up in brown paper and containing another $200,000 had not been taken and was still lying on the table next to the now destroyed safe.

Well, friend, the alarm was given and the telegraph wires flashed the news everywhere. This, of course, caused the newspapermen to smell a sensational scandal and by noon on Friday not only was the Oil Region abuzz with talk of the robbery, but the rest of the United States knew about it as well. Headlines from one side of the country to the other screamed out about "The Great Benninghoff Robbery," and the news was the talk of the day.

After they had left the Benninghoff farm, Saeger and his cohorts drove to Louis Wardle's hotel to divide up the spoils before heading their separate ways. The bulk of the spoils went to Saeger, Elliott and McDonald and a smaller share went to both Geiger and Shoppert for their parts in the night's events. Nor, did they forget Wardle, the hotelier; he was given $1,300 to ensure his silence in the matter. The loot being divided, the two thieves from Baltimore hung around for about a week before heading to Canada. Saeger went back to his home and business and Shoppert and Geiger returned to their lives. As far as they were concerned, they had gotten away with the robbery. Alas, this was not to be.

The countryside soon was swarming with detectives eager to make a name for themselves or to reap the reward of $10,000 that was being offered. Among all of these, the ex-chief of police from Pittsburgh was perhaps the most active and alert. His name was Hague, and he was the one who broke the case after three months of work. It happened after George Miller, who had nothing to do with the robbery, got into an argument with Saeger over a poker game. It was during this argument that Miller blurted out that he knew all about the Benninghoff Robbery. In order to pacify him and keep him quiet, Saeger paid him $1,000.

Unfortunately, that was the mistake that broke the case wide open. For you see, Miller was friends with Jacob Shoppert, who had taken part in the robbery, and together the two of them made the rounds spending money in a manner that caused Hague to put a tail on the both of them. It was a good thing, at that, for while Shoppert was visiting a town on the edge of Ohio, he was arrested, and while in the confines of the jail he asked for pen and paper. It was on this paper that he wrote a letter to Wardle, reproaching him for not sending the money to bail him out of jail. Wardle by this time had sold his hotel in Saegertown and moved to Ohio to start a brewery. The letter, which the jailers gave to the detectives, was enough to arrest Miller, Shoppert and Wardle for the Benninghoff Robbery. The three were tried, convicted and spent a short term in the penitentiary. Geiger was not tried because no one could prove beyond a shadow of a doubt that he had been part of the plan. McDonald and Elliott were captured in Toronto, but their Canadian lawyers were able to find enough flaws in the paperwork that Hague carried to keep them from being extradited.

All this leaves Saeger, the fellar who thought up the robbery in the first place, to be accounted for. He was not suspected until after he had sold his business and left for the West. Hague lost sight of him for three years, when just perchance a noted cattle king of the Texas–Colorado trail happened to enter a saloon in Denver, Colorado, with several of his friends. Well, sir, behind the bar that day was Gus Peiflee, who had come west from Meadville,

Pennsylvania, and he recognized this cattle king as none other than Jim Saeger. It was Peiflee who wired back East to inform them of the fact that Saeger was now in Denver. In very short order, Chief of Police Rouse and John Benninghoff himself arrived in Denver with extradition papers for Saeger and arrested him. Saeger coolly told the both of them, "You'll be a devilish sight older before you see me in Pennsylvania." Rouse and Benninghoff were then informed that over a hundred of Saeger's cowboys were in Denver and they were reckless, lawless fellows who were certain to kill anyone who attempted to take their boss away. It was because of this that Rouse and Bennighoff dropped the matter and returned to the Oil Region alone. Saegar stayed on in Texas, where he was respected and prospered. It was said that he was fair and just in his dealings and that not long after the death of her husband, he sent $5,000 to the widow of George Miller.

Well, friend, that is the tale of the Benninghoff Robbery. At the time, you can be sure that it attracted more attention than the sight of the first woman to wear bloomers.

A SELF-DRILLING WELL

One of the strangest tales ever to come out of the Oil Region occurred just outside of Rouseville, Pennsylvania, in 1870. The outcome was portended by the combination of petroleum and natural gas in the same well.

Now, natural gas is the cleanest and handiest fuel ever to be used to warm hearth and home. However, when one is drilling for it, it can be deadly. The reason is simple: all it takes is one spark to ignite natural gas and cause a very large explosion.

This is what happened to one well outside of Rouseville. The men engaged in drilling the well had unknowingly drilled down into a pocket of natural gas. Normally this would not be too much of a problem, but for some reason their bit struck just the right stone at just the right angle and caused a spark that ignited the pocket of natural gas and caused a large explosion.

The reason that this happened has to do with the way one drills using the bit-and-string method of drilling. In this method, the drilling bit is a large, round piece of iron or steel with one end flattened out and a hole drilled in the other end. The flattened end is then further shaped into the form of a wedge. It looks very much like a large, wide pry bar or spud bar. It is the job of the tool dresser to make sure that the bit stays sharp and ready to use.

When the drilling rig and engine house have been constructed and set up, the first thing to happen is that several lengths of casing are driven down through the soil until bedrock is reached. Then, the driller attaches the bit to the end of the drill string. The other end of the drill string is attached to the walking beam. As the engine runs and powers the band wheel, it rotates, causing the walking beam to pivot up and down. This up-and-down motion is then transferred through the drill string to the drill bit at the bottom of the well.

As the bit rises and falls, it chips away at the material beneath it. The driller has to keep turning the drill string cable so that the square bit he has down

An oil rig similar to the one that was burned in 1870 outside of Rouseville, Pennsylvania.

at the bottom of the hole will turn and drill the desired round hole through the ground.

Now, this was just what was happening outside Rouseville in 1870, when they hit that pocket of natural gas. Once the natural gas had exploded, it sought to expand as all exploding gases expand. Unfortunately for the driller and tool dresser, the only route for expansion was straight up the well hole. Try to image the scene, friend: there you are at work drilling a well when suddenly you hear a muffled *whomp* that is followed by a *whooshing* sound, and the next thing you know, a jet of flame is shooting up out of the well that you've just been drilling.

Well, sir, this is what happened then and even more so, because that jet of flame was so intense that it caught the drilling rig on fire. The rig had been constructed out of wood and wood likes nothing so much as to burn when given half a chance and a flame to get things going. All things considered, it was a wonder that the only damage done that day was material. The drilling rig and engine had burned to the ground, and after the initial fire, the well had settled back down to being another hole in the ground.

At this point, the producer could have done one of two things: first, he could have walked away and written the well off as a loss; or second, he could have rebuilt the drilling rig and engine house and then figured out a way to get things back on track and make the well a producing venture. As it happened, he chose the latter and set his men to rebuilding the rig and buildings at the wellhead.

Here we get to the strange part of this tale, for as the men were working to rebuild the buildings and the drilling rig, they could hear the natural gas building up underneath the drilling bit and cable string that were still down the well. Then, every so often, enough of the gas would have built up to lift the drill bit up toward the surface. When the drill bit was high enough, the natural gas would escape around it and the bit would fall back to the bottom of the well and the process would start all over again.

This strange occurrence kept up right until the day that everything was ready for the men to resume drilling the well. The first order of business was to get the drill string out of the well. To accomplish that feat, they hooked a set of fishing jars to the end of the drilling cable and lowered them down the well to fish the drill bit out. The drill bit was down farther than they thought, so it took more than a few tries, but they were eventually successful in bringing the drill bit back to the surface.

As soon as the drill bit was on the surface, the tool dresser took charge of it and began reshaping it so that it could be used to finish drilling the well. After he was done, he helped the driller hook the bit onto the end of the drill string, and as they began to run it down the well, they noticed something odd in their well. Instead of the bottom of the well being way down deep inside the well, it looked close enough to touch with a long stick.

The driller found a long stick, and as he ran it down the well, he could tell that the bottom of the well was not only close but penetrable. When he pulled the stick back out, much to his amazement, it was covered in petroleum.

During the time that they had been rebuilding the drilling rig and engine house, the constant accumulation and release of natural gas below the drill bit had caused the well to drill itself down into a petroleum-bearing region and the natural gas had further provided enough pressure to bring the petroleum to the surface once the plug that the drill bit had created was removed.

That fact makes it one of the strangest wells to have been drilled in the Oil Region.

STANDARD OIL AND THE SOUTH IMPROVEMENT COMPANY

Perhaps no other company has been more maligned and misrepresented than the Standard Oil Company. There was a time in the Oil Region when if a man drilled a dry hole, chipped a bit, broke a cable or went bust on anything, it was the fashion of the day to blame it all on Standard Oil.

There are many reasons for this. For one thing, Standard Oil was rolled into the same lot as the South Improvement Company even though, while they did have major shareholders in common, the South Improvement Company was not a Standard Oil brainchild. Also, envy was a big part of the reason as well. If someone does a job better than anyone else and makes money at it, they are sure to be envied in their success. Lastly, Standard Oil was bringing a new style of doing business to the Oil Region and anything new is always met with hostility.

The initial concepts behind Standard Oil had been tried before in the Oil Region, but it was in 1872, with the formation of the Standard Oil Company of Ohio, that they finally took hold. In the beginning, the company was formed as a way to overcome the problems caused by waste in the petroleum industry.

For you see, prior to 1872 about two-thirds of the crude petroleum that was pumped from the ground was lost, either due to spillage during transportation or waste during the refining process. This made it very hard to turn a profit, even with the market prices as high as they were. It was the goal of Standard Oil to rectify this matter.

To that end, they began by pooling the resources of several refineries around Cleveland, Ohio, in order to get a better rate for shipment from the railroads. It worked like this: because there were several refineries

John D. Rockefeller, the founder of the Standard Oil Company and the South Improvement Company.

buying their petroleum as a block, they could negotiate a lower rate for shipment because of the amount they were having shipped. Then, when they purchased their own tank cars and began loading and unloading them on their own, they were able to get even lower rates from the railroads and eliminate the cost of the barrels required to ship from the wellhead to the refinery.

The next big step that they took was to improve the refining process. They did this by bringing in chemists and mechanical experts to take a good look at the process and eliminate the waste products from the process. To that end, they were so successful that not only was the quality of the end product increased, but the retail cost came down as well.

Friend, it is true that this advancement fostered by Standard Oil caused many of the older refineries to close down because they could not afford to make the needed improvements. Still, other refineries joined with Standard Oil to become part of the greater company and others were able to make the necessary improvements and remain independent. In the long run, this improvement of the refining process turned out to be one of the best things that could have happened to the petroleum industry. Progress such as this had taken place before in other areas of the industry and it would certainly happen again.

The final significant change that Standard Oil was able to make was to put better control on the initial production of crude oil right out of the ground. Drillers, trying to make a quick dollar by putting down more wells, had put more petroleum on the market than was really needed. This action, instead of making the drillers rich, had the effect of driving down the price of oil. When Standard Oil stepped onto the scene, it began putting the excess petroleum into large storage tanks until the time it would be needed.

So it was that on January 2, 1882, forty people came together for a meeting. They were all associated with Standard Oil and owned the entire stock of fifteen corporations and part of the stock of a number of others. Now, nine of these people controlled the majority of all of those shares. It was during this meeting that they decided to combine all the stock into a single trust and make the nine who controlled the majority the trustees so that they could conduct the business of all the corporations with the best interests of all concerned. This was the inception of the Standard Oil Trust.

In the ten years the trust was in existence, it went from forty stockholders to two thousand. Any employee could save up his money, invest in the trust and become a stockholder. However, at no time did the trust ever work to corner the market, and each company that made up the trust was managed as an individual company. The exception was that if one came up with an idea for improvement, that idea was passed along to all the others.

On the other hand the South Improvement Company was formed in 1871 by a group of major railroad interests led by Thomas Scott of the Pennsylvania Railroad. The main goal was to set up a secret alliance between the major railroads and a select group of large refineries.

The way the alliance was to work was that the railroads would raise the rate they charged to transport petroleum from the wellhead to the refinery. This rate would be paid by everyone, but those refineries who were part

The Standard Oil building located at 26 Broadway in New York City, circa the late 1800s.

of the alliance would receive up to 40 percent in the form of a rebate or a kickback from the railroad. In addition to the kickback that the member refineries would receive, they would receive a portion of the amount that was paid by nonmember refineries as well.

John D. Rockefeller was brought into the plan because he had merged several Cleveland, Ohio refineries and had started into his own negotiations with three of the major railroads that ran through Cleveland. These railroads were the Erie, the Pennsylvania and the New York Central.

It 1872, when word leaked out about the alliance and the planned increase in the rail shipping rates, the independent producers and smaller refineries got up in arms and marched on the railheads. This outrage became known as the Oil Riots of 1872 or the 1872 Oil War, although it did not quite escalate to that level.

Even though the railroads agreed to back down and treat everyone the same in the future, the Pennsylvania state legislature repealed the charter of the South Improvement Company. At this news, the independent producers celebrated a victory, but Rockefeller thought it was ridiculous to call the company a conspiracy as many had. He would later argue that rebates had been a common business practice before the South Improvement Company used them, and they certainly continued to be afterward as well.

THE OIL RIOTS OF 1872

A round the end of 1871, rumors began to circulate throughout the Oil Region about a new company, the South Improvement Company, and its plans for the petroleum industry. These rumors included the fact that the amount the railroads were charging to transport petroleum would be rising by almost 100 percent. However, it was not until February 1872 that the rumors were confirmed and the people of the Oil Region began to realize that while this plan was being touted as for the good of the industry, it would be anything but a benefit to the industry.

The increase in the rate to ship petroleum would mean that not only would it cost more to make the shipment, but also the amount that each producer would make in profit would go down. This could have the effect of ruining many producers, if not the entire Oil Region itself. This information was bad enough, but it got worse when it was revealed that a new company, the South Improvement Company, would not be affected by the new rates. In fact, it would get a special rate all its own.

As the facts started coming in, the rumor proved to have more truth in it than not. The shipping rate from the Oil Region to Cleveland, Ohio, jumped from $0.35 a barrel to $0.65 a barrel at the same time as the rates to Buffalo, New York, jumped from $0.40 to $0.65 a barrel. Then, not a week later, the rate from Warren, Pennsylvania, to New York City was raised from $0.87 a barrel to $2.14 a barrel.

It was on the morning of February 26, 1872, that the producers, drillers and other oilmen found out that the much-threatened rise in the rate of transportation had not only come true, but the costs also had to be paid in advance. As if that were not enough, it was also made publicly known that the South Improvement Company had been given its own special rate and did not have to pay in advance. This news spread throughout the Oil Region

An oilman upset over hearing about the new rules being imposed on the oil industry by the South Improvement Company.

and caused everyone to take to the streets to find out what their neighbors thought about the matter.

Just a mere twenty-four hours later, there was a gathering of over three thousand very excited oilmen at the Opera House in Titusville. One of the main thoughts that was passing from lip to lip was that this was some sort of a conspiracy to put them all out of business. Everyone in the crowd seemed to be of one mind, be they driller, pumper, refiner, broker or producer.

Three days after the Titusville meeting, an even larger meeting was held in Oil City. The temper there was even more combative. Together, they organized a Petroleum Producers' Union and agreed not to start any new wells for the next sixty days, thereby reducing production. To further reduce production, they agreed not to pump on Sundays and not to sell to anyone known to be a part of the South Improvement Company. It was also put forward as a good idea to support those refineries and railroads that had refused to join this new conspiracy. Lastly, they agreed to start building their own lines over which they would have complete and total control.

At the end of this meeting they sent a committee to the state legislature asking that the charter of the South Improvement Company be removed. Another group went to Washington, D.C., to see if it could get an investigation started as to whether the South Improvement Company was interfering with the free flow of goods and services.

While this was going on, the Petroleum Producers' Union had a history of this perceived conspiracy written up, listing the names of those involved and the plans of the company for the Oil Region. They printed over thirty thousand copies of this history and sent it out to every judge they could find, along with all of the United States senators and members of Congress. Copies of the history also were handed to affected state legislatures, all railroad men and many of the prominent businessmen in the country.

By now, it should be becoming apparent that these were some determined oilmen, and stubborn as well. With everything else going on, the oilmen decided to reintroduce into the state legislature a ninety-three-page-long petition demanding a free pipeline bill.

Friend, the long and the short of it is that for weeks the men of the Oil Region abandoned their regular business and surged from town to town, intent on bringing down the South Improvement Company. However, the curious thing in all of this is that very little, if anything, was said against the railroads. If you were to have questioned the men leading this charge about that fact, they would have told you that they expected to be cheated by the railroads. However, they drew the line when men from their own industry started cheating them.

Things went from bad to worse the more they found out about the South Improvement Company. Things got really bad when the news came out that the company was a secret organization and had been at work under their very noses for several weeks. This bit of news so alarmed one group of oilmen that they started their own investigations. Unfortunately, they were not able to learn anything until they agreed not to reveal what they'd learned.

The most shocking discovery of these investigations was that somehow the members of the South Improvement Company had convinced the railroads that they in fact not only owned, but were also the spokesmen for the entire

Oil Region and that, hereafter, no petroleum would be shipped unless they shipped it.

Even after everything had come to light, most of the oilmen still wanted to know who made up the South Improvement Company and whether its members considered themselves conspirators or not. Once the names of the individual companies were known, they were boycotted by the members of the Petroleum Producers' Union.

The boycott went so far as refusing to deal with any refinery on the list, even though the men who bought petroleum for the refineries were young and well liked and offered to purchase crude petroleum at anywhere from seventy-five cents to a dollar more than the market price. It was simply not sold to them. In fact, friend, at one meeting they were ordered "to desist from their nefarious business or leave the Oil Region."

In the end, the Petroleum Producers' Union defeated the South Improvement Company and the Pennsylvania legislature pulled its charter late in 1872. The life of the company had ended before it had made even one valid transaction.

THE TRUE STORY OF
RIAL & COMPANY'S NO. 9 WELL

In northwestern Pennsylvania, where French Creek joins the Allegheny River, you'll find Franklin, the county seat of Venango. Now, just across French Creek from Franklin is a very large hill. A little over a hundred years ago, this hill was called Point Bluff and it is upon this hill where our tale takes place.

It was during the summer of 1881 that the men working for Rial & Company of Franklin began drilling a well on the top of Point Bluff. This was the ninth well that they had drilled, and as such, it was called The Rial & Company No. 9 Well. Oilmen can be a practical lot when it comes to naming things.

All went well until November 23. Now, that day began just like any other day for the men drilling the well. Yet during the morning hours, just as they reached the 300-foot mark, the drilling cable jacked tight. This meant that they had drilled into some sort of a cave or cavern, or a drift. Now, friend, you need to remember that this was only twenty-two years after Col. Drake had made his strike up near Titusville, and every man on the crew knew that Uncle Billy Smith had experienced just such a drop the day before his strike. With this in mind, the rope choker tied a small cord onto the drill string and then let the drill string play out slow and careful like so that they could measure on the way down. Those tools went down another 8 feet before they bottomed out. At this point, they pulled that tool string up all 308 feet just as slow and careful like because they wanted to see if anything was on the tools. If they had struck oil, their work was over and they could be paid off and move onto the next well after a grand night at the local thirst parlor.

The house of Edwin Rial in Franklin, Pennsylvania.

Truth be told, friend, there was a liquid on the tool string that day. However, this liquid caused a mighty consternation among the men at the wellhead. The reason being that, the liquid just didn't look like oil, it didn't smell like oil and when they reached out and felt it, it didn't feel like oil. Well, the crew didn't know what to make of it, nor what to do, so they decided to go and get the boss. It was old Edwin Rial himself who came up the hill that day, and he had them run the bailer down the well to see how much of this strange-looking liquid there was down there. Now a bailer, sometimes called a slush bucket, is nothing more than a big, long tube with a plug on the bottom of it that is used for getting slush out of the bottom of a well. Slush is a combination of water, gas, oil, ground-up rocks, salt water, sand and other things that happened to be drilled through on the way down. It's a very slimy, grimy mess. Well, friends, that day, that slush bucket came up full of this strange-looking liquid.

Now, the men still didn't know what it was. It didn't look like oil. It didn't feel like oil. It didn't smell like oil, and when they drew a mug of it off and sloshed it around, it sounded like any other liquid. At that point they were in a quandary over what to do. The only thing left to do was to taste it and no man at the wellhead, Rial included, wanted to be the first to taste a strange liquid that had just come up 308 feet out of the ground.

Well, the mug was passed from man to man until finally it reached a young lad of ten or twelve years of age. He was there to learn the drilling trade and was the low man on the totem pole. Of course, he didn't want to taste it either, but he did want his job. So, he very shakily raised that mug to his lips and took a small taste. He was so astonished by the taste that he cried out, "This tastes like beer!" and proceeded to gulp down what was left in the mug. The other men at the wellhead tasted it for themselves and declared that it was some of the finest tasting beer they had ever had.

Unfortunately, they now had another problem. For you see, we've struck water and salt water, oil and natural gas, and they say that there are wells in Australia that you can get opals out of; however, nobody had ever struck beer before and if word of their strike ever got out, every wildcatter, slant-well driller and spring pole man from miles around would be crowding in on their strike. After much thinking, they hit upon a solution. Sometimes there is too much liquid and slush at the bottom of a well to get it out with a slush bucket. When this occurs while drilling a well, the drilling rig can be switched over to pump the stuff out. This is known as pumping down a well, and for all intents and purposes, it looks like the well is still being drilled.

So it was that in very short order the men had switched over from drilling to pumping, and they commenced to begin enjoying their good fortune. Now, accounts of how long this party lasted vary anywhere from a couple of hours to a couple of weeks, but ultimately it was too good to be true.

Now, friend, what happened was that several years earlier the Grossman Brewery, also of Franklin, Pennsylvania, had dug a large vault into the base of Point Bluff. They had done this to get an even temperature for aging and fermenting their beer. While we did have mechanical refrigeration at the time, it was very expensive and not widely used. Basically, one had two choices for keeping things at an even temperature: cover them with snow and ice or stick them in a hole in the ground. Phillip Grossman, who owned the brewery, was in business to sell his beer, not to keep the ice merchants in business, so he stuck his beer in a hole in the ground.

Now, just perchance, Rial & Company No. 9 Well had come down smack dab right in the middle of the largest aging vat of the Grossman Brewery. The way that Grossman found out about all this was that a young lad, again of ten to twelve years of age, who was working for him and learning the brewing trade, was checking the aging vats. In the course of this, he noticed that the level in the largest vat was down quite considerably from where it should have been.

Now, friend, it was Phillip Grossman himself who thundered up Point Bluff that day to see just what was happening to his beer. After finding out what was going on, the last the men at the wellhead saw of him that day was his back as he thundered off toward the offices of Rial & Company to give Edwin Rial a good piece of what for.

Friends, you have to understand that Rial owned the mineral rights and was perfectly within his legal rights to drill where he had been drilling. He was just not so within his legal rights to be taking out Grossman's beer. Therefore, he had his men lay off while Phillip Grossman moved his aging vat and salvaged what was left of his beer. They then continued drilling about another forty feet and struck oil at around fifty barrels a day. That was not a bad well for that region and that time.

At this point in this tale, most people have but one question in their minds: "Did they get to drink the beer for free?" The answer to that is unfortunately and emphatically, "No, they did not." For you see, friends, about a month after all this transpired, a bill showed up at the offices of Rial and Company from the Grossman Brewery, and legend has it that one day's production at the then-current prices paid for all of the beer that was drunk out of the "World's Only Beer-Producing Well."

THE GREAT FLOOD
AND FIRE

It began with a severe thunderstorm in June 1892. As the rain kept coming, most people along Oil Creek knew to expect a flood. However, when lightning struck a stock tank at the Crescent Oil Refinery Company, the resulting disaster was more than most people could have imagined.

Friend, the area around Oil Creek has always been prone to flooding. So much so, that the people who lived there developed a way to make artificial floods for the use of moving cargo downstream to Oil City and beyond. Thus, when the rain started up again that week and continued for several days, most of the people who had lived along Oil Creek for any length of time started getting ready for another flood.

It had been raining more or less constantly for the better end of a month, and all the small streams that fed into Oil Creek were already flowing at capacity. The additional rain caused them to start flowing over their banks.

Seven miles above Titusville, at the town of Spartensburg lay the Thompson & Eldred Dam. Behind this dam was a lake that was a little over one and a half miles long and a quarter mile wide. The constant rains had put enough extra water into that lake so that as midnight approached, it was close to bursting. It was during the first few minutes of June 5, 1892, that the dam finally gave way and sent all that water rushing down Oil Creek.

Reports from Titusville after the fact tell how half of the town was overrun with the rising floodwaters, which washed away many of the smaller buildings and caught so many people unaware that they were swept away as well. When the flood had subsided, almost a third of Titusville was lost and of those people who were swept away, seventy-five were known to have died.

Flames leaping up from an oil well.

It is assumed that the rest were able to get to shore farther downstream and find their way back to the city.

As the waters of Oil Creek claimed the lower parts of the city, brave men took to boats to offer what aid they could. All through the night, calls for aid and screams of pain could be heard coming from rooftops, tall poles and other debris that was above water level. It was through the work of these men that hundreds of people were brought safely out of the floodwaters.

At one point, five men who were all clinging to the same floating tree vanished beneath the rising waters. It is estimated that somewhere between one hundred and two hundred people were carried off that night.

As the floodwaters were reaching their crest and the people began to think that the worst was over, lightning struck the nearby Crescent Oil Refinery. It is hard to say whether it hit the stock tank exactly or just close enough to ignite it. Either way, the result was the same, and soon the refinery was ablaze.

It only took moments before the fire was climbing almost two hundred feet into the sky, and because the refinery was flooded, the fire was able to travel with the petroleum that was floating on the water from building to building, gaining in strength. So it was, friend, that in less than three minutes' time, the entire refinery was engulfed in flames.

This led to the inevitable panic as people realized what was happening and did their best to get as far away as possible. While the fire put a lot of smoke into the air, it also brightened the night enough so that those who were fleeing had no trouble seeing where they were going.

It only took an hour for the fire to spread and cover an area of several acres. There were two main reasons for this, friend: first, almost every building in Titusville had been made of wood and, soaked or not, wood burns; and second, there was a great deal of petroleum soon floating on top of the floodwaters. Even in its raw, crude form, petroleum burns quite readily. There is always some petroleum floating on the waters of Oil Creek and the floodwaters had knocked over many a stock tank, adding to this supply.

By this point, the town waterworks, the electric plant and the illumination gasworks were all under water, leaving the residents of Titusville without their accustomed sources of fuel, water and light. The fortunate thing was that the flow of natural gas to the illumination works had been shut off when the water levels began to rise. Had it not, the disaster could have been much worse.

That night, Oil Creek had swollen to five hundred times its normal size and was full of all kinds of flotsam and jetsam, some of it alive and much of it not. Scattered throughout this floating, burning morass were hundreds of people who had been swept away from their homes in the middle of the night and were now clinging to whatever they could find that would float.

The bodies, when found, were taken to temporary morgues that were set up at the undertakers Davidson & McNitt. Most of the bodies showed signs of having been burned, either as the cause of death or after they had died. In many cases it was extremely hard to say who these people had been in life and in a few cases, it was impossible.

A pumper watching his well pump petroleum into a stock tank just days before the Great Flood and Fire.

As the morning wore on and as the enormity of what had happened started to sink in for those who had survived, people banded together to aid each other. The Rouse Armory was turned into a relief center and temporary hospital, where over one hundred people were cared for. To make matters even worse, there was not a bridge for miles around that wasn't damaged or washed away in the flood.

If things were bad in Titusville, even worse was in store for Oil City. The flooding there had not begun until the late morning hours of June 5, 1892.

It was nigh on eleven thirty when the floodwaters really started rising. By then, most of the populous had gathered along the banks of both Oil Creek and the Allegheny River to watch the rising waters and see just how high this flood would get.

It was around the time when the main crest of the flood came roaring into the valley that the residents realized it was covered in petroleum. This knowledge was foreshadowed by the clouds of distillate gas and benzene that could be seen above the surface of Oil Creek. Because of this, the people on the bridges began to climb back to higher ground. However, no sooner had they begun to do this than an explosion was heard upstream. The first explosion was followed by two others. These three explosions ignited the petroleum that was floating on the water and the gases that were hanging in the air over Oil Creek, turning the scene into a boiling, rolling mass of smoke and flames.

Oil City was built where Oil Creek empties into the Allegheny River, and over the many years since the last ice age, the two have carved the land down into a valley. It was this valley that kept the fires contained to Oil City and did not allow them to spread out over the land as they had done in Titusville. The portion of the city through which Oil Creek flows is known as the Third Ward, and in under three minutes, it was engulfed in flames. While no one is exactly sure how many people died that morning, the number certainly is in the hundreds.

In the ensuing mayhem, people forgot who they were and became a frightened mob trying to get to the high ground. It was not until they reached the higher levels that they once again became civilized people and began to worry about what had happened to their friends and relatives.

The entirety of North Seneca Street was engulfed first by the floodwaters and then by the fire. Many of those who had climbed to the top of their houses in order to escape the floodwaters found themselves diving into these self-same waters in order to escape the encroaching fire.

The damages to Oil City and Titusville were immense and were said to have reached over $1.5 million for each city. Friend, this amount is even larger when you take into account the fact that, at the time, the combined population of the two cities was only a little over twenty thousand people. Then, too, there is the eighteen-mile stretch of Oil Creek that lies between the two cites. That area was devastated as well. In total, it is believed that somewhere around three hundred to four hundred people lost their lives during this disaster and the damage totaled over $3 million. Additionally, some thirty miles of creek banks were laid to waste during the course of the Great Flood and Fire of 1892—truly one of the worst disasters to ever hit the Oil Region.

SPINDLETOP

It was in the oil fields of Texas in 1901 that the death knell sounded for the Pennsylvania Oil Region. Spindletop is a small hill or mound that sits just outside of Beaumont, Texas. Originally, it was called the Sour Springs Mound because of the fact that natural gas seeped out from under it and into the surrounding water table. Now, friend, the geologists tell us that this mound had been formed by a gigantic underground dome of salt. Now, at some point this dome of salt began to move toward the surface of the earth and in doing so pushed the soil that was in its way upward as well.

The local Native Americans had known about the oil seeps that could be found in and around East Texas for many, many moons. They, like those from the Seneca Nations, had used the tarlike petroleum that they found at these seeps to treat a vast number of ailments, even to the point of drinking it in order to cure internal ailments. Then, in 1543, the Spanish explorers found a black, sticky, tarlike substance on some of the East Texas beaches and used it to waterproof their equipment.

The long and the short of it is that the Texans knew they had petroleum under their feet, and it was Lyne T. Barret who sunk the first petroleum well near Nacogdoches in 1866. This field, which was known as Oil Springs, would not be fully exploited until 1888, when a crew of drillers came down from Pennsylvania to show the Texans how it was done. One of the first wells they sank came in at a rate of between 250 to 300 barrels a day. Word of this size of a strike was sure to attract other producers and drillers to the field. True to form, it did just that, and it was only a matter of time before the true potential of the Texas Fields was realized.

The first big oil field to be opened up was the Corsicana Field in East Texas. Like so many other fields before, this one was discovered by local businessmen who sank down a couple of really deep wells looking for water.

A gusher from the Texas Oil Fields.

When they hit the petroleum sands, they often couldn't see past the noses on their faces and kept drilling right on through those petroleum sands in order to find the water they were looking for. However, H.G. Damon and Ralph Beaton saw this black gold mine for what it was and quickly formed the Corsicana Oil Development Company.

John Galey, a Pennsylvania driller, was brought in and the team drilled a couple of marginally successful oil wells in 1896. Because these wells all flowed at twenty-five barrels a day or less, Galey and his partner, James Guffy, sold their interest in the Corsicana Field and headed back East, convinced that there was little to no future in the Texas Fields.

The locals had more belief in Texas Petroleum, and by the end of 1900, they had produced more than two million barrels of petroleum from the Corsicana Fields alone. While this was not a large number by the standards of the Pennsylvania Fields, it certainly showed that there was petroleum under Texas.

It was Patillo Higgins, a one-armed mechanic and self-taught geologist, who believed that the future fuel of the modern world would be based on petroleum and not coal. The question then became where to find that petroleum. Somehow, he became convinced that it was just waiting for him underneath the ground at Spindletop; he just had a feeling that drilling down through the top of this salt dome and others like it would produce dramatic results.

In 1892, he succeeded in generating enough local capital to fund the exploration of Spindletop, and he organized the Gladys City Oil, Gas and Manufacturing Company. It made three attempts at drilling a well down into the mound. Unfortunately, all three failed due to a large layer of quicksand, and the company's funds were nearing exhaustion. It was then that Higgins struck a deal with Captain A.E. Lucas.

Higgins met Lucas because of an ad Higgins had run in a local newspaper. In fact, Lucas was the only one who answered the ad. Now, he was a real captain, having served in the Austrian navy, and he was a trained engineer and experienced salt miner. Unfortunately, the first wells he drilled with Higgins were little more than dry holes, and the money was running thin.

At the end of his rope, Lucas contacted both Guffey and Gailey, both of whom had left the area three years earlier. While they had been unconvinced on the subject of Texas Petroleum, they were willing to return and give it another shot. It was to that end that John Galey returned to Beaumont in 1900 to survey the area. He picked a spot, and the drilling began on October 27, 1900.

Galey had suggested to Lucas that he hire the Hamill Brothers to drill the well, and it turned out to be a good thing he did as the drilling was difficult from the get-go. Lucas and his man ran into a common problem when drilling in East Texas. The problem was and is that there is no real rock surface to spud down to. Instead, one had to drill through several hundred feet of sand. Doing this made the new wells prone to caving in.

It was Curt Hamill who came up with the answer. The answer was quite revolutionary for the time and it changed the way wells have been drilled ever since. Hamill figured that if he pumped mud, instead of water, down his rotary drilling rig to clean out the cuttings from the business end of things, it just might help stick things together as they continued drilling down into the earth. It was much to everyone's pleasant surprise that the mud mixture Hamill came up with not only did just what he thought it would, but it also did more. That is why mud has been used on just about every rotary hole drilled around the world since that time.

Two months of very hard, exhaustive drilling had brought them to a depth of 880 feet. Since it was Christmastime, they took the week off and planned to start again after the New Year. It was on New Year's Day 1901 that everyone arrived back at the wellhead, energized and ready to get back at it. They were moving right along when, a week later, at a depth of 1,020 feet, they had to pull the drill string out to change some equipment.

On January 10, 1901, they began running their tool string back down the hole. When they reached the seven-hundred-foot level, the mud started bubbling back up out of the hole and then their tool string shot up into the air with great force. This was followed by silence; nothing else seemed to be happening.

After giving the well a few minutes as they talked over what could have happened, the men got back to work. The first task was to clean up the mess and see what could be salvaged. They were involved in this task when a sound like a cannon shot came from the open well. This was followed by a stream of mud that blasted out of the well like a rocket. A few seconds after that, the men were surprised to see natural gas shooting up into the air. However, their surprise didn't last long, as petroleum followed the natural gas in very short order.

This oil gusher rose to a height of more than 150 feet into the air, something close to twice the height of their drilling rig. When Captain Lucas began this well, he was hoping for an initial production of five or ten barrels a day. Lucas No. 1, as the well was now called, came in at nearly a hundred thousand barrels a day, which was a higher rate of production than the combined production of all of the other wells in the country at the time.

In very short order, the population of Beaumont swelled from ten thousand to fifty thousand. Not only that, but before the end of the year there were over two hundred producing wells on Spindletop, and they were owned by over a hundred different petroleum companies. The gusher and oil field at Spindletop was the beginning for many companies that were to become giants in the petroleum industry.

It was the beginning of a new era in the petroleum industry, one that would sadly leave the Pennsylvania Oil Region a forgotten piece of history.

OIL HISTORY TIMELINE

Since this is a book about history, I felt it only proper to include a timeline within its covers. The events and dates listed here are only a small part of the historical events that occurred in and around the Oil Region. I offer these as a guide not only to give you an idea of what happened, but also as a starting point from which you might dig further on your own and find out things that you never knew, but which affect your life every day.

A smart operator will notice that I've ended this timeline with the strike at Spindletop. By this, I am not trying to say that the strike at Spindeltop was the end of the petroleum industry. Far from it: the industry has only continued to grow and prosper. However, it was that strike that shifted the focus of the industry firmly and finally away from the Pennsylvania Fields and onto other locations around the globe. While there is still a remnant of the industry left here, it is but a ghost of its former self.

1748	A map is published in Sweden by Peter Kalm showing oil springs along Oil Creek in Pennsylvania.
1767	The practice of retrieving oil from the top of creeks and rivers by skimming or soaking blankets and wringing them out by members of the Haudenosaunee Nation is recorded by Sir William Johnson of New York.
1778	Moravian missionaries in Western New York report that members of the Seneca Nation used the products of "oil wells to carry on trade with members of the Niagara Nation."

A drill house and rig tower from along Oil Creek.

1785   General William Irvine notes that due to an oil or bituminous matter that floats on its surface, there is a stream in western Pennsylvania named Oil Creek.

1790   Nathaniel Carey of Titusville, Pennsylvania, begins skimming oil from nearby springs and delivering it to customers by horseback.

1791   A map of Pennsylvania shows a stream named "Oyl Creek."

1806   The drilling of a salt well in West Virginia is ruined when oil comes up along with the salt water.

1840   Natural gas is first used in the production of salt through brine evaporation outside of Centerville, Pennsylvania.

1840s   Oil begins appearing in the salt wells of the Kier Family of Tarentum, Pennsylvania.

1846   Samuel Kier begins to market Kier's Petroleum Butter, shipping it as far away as California.

1848   Samuel Kier begins experiments to refine petroleum for illumination purposes.

1849   Samuel Kier begins to market Kier's Medicinal Petroleum and establishes the world's first petroleum refinery in Pittsburgh, Pennsylvania. It has a five-barrel still.

1853   The first written document looking into a mechanical development of petroleum production is signed by J.D. Angier of Cherrytree Township, Venango County, Pennsylvania, and the firm of Brewer, Watson & Co. The firm of Brewer, Watson & Co. was primarily associated with an extensive lumbering business along Oil Creek in western Pennsylvania.

1854   In a business deal, 105 acres of the Hibbard Farm are sold to George Bissell and Jonathon Eveleth, who then organize the Pennsylvania Rock Oil Company, incorporating in New York, with the purpose of making petroleum a commercial product.

1855   Yale Professor Benjamin Silliman Jr. completes a thorough analysis of petroleum taken from a spring on Oil Creek.

1858   In New Haven, Connecticut, the Seneca Oil Company is formed out of the remnants of the Pennsylvania Rock Oil Company, and Edwin L. Drake is hired to oversee the drilling of an oil well on the Hibbard Farm.

1859   On August 27, at a depth of 69.5 feet, William "Uncle Billy" Smith, drilling for Drake and the Seneca Oil Company, strikes oil, becoming the first man who had deliberately set out to drill for oil to do so.

On August 30, John Gradin and H.H. Dennis drill a well in Tionesta, Pennsylvania, that fails to strike oil. This is the first dry hole of the industry.

On October 7, a spark ignites natural gas fumes and the resulting fire burns the buildings at Drake's Well to the ground. This is the first fire at an oil well. The buildings are rebuilt and production continues.

1860   The first load of petroleum is carried from Oil Creek to Pittsburgh aboard the steamship *Venango*.

1861   In October, The Phillips No. 2 Well on the Tarr Farm comes in at four thousand barrels a day. This is the industry's first gusher.

In November, the first shipment of petroleum crosses the Atlantic Ocean, traveling from Philadelphia to London, England.

Oil History Timeline

1863   The first antipollution bill is passed by the Pennsylvania legislature. Its aim is to prevent tar and other liquids from flowing into certain creeks and rivers.

1865   On January 7, The Frazer Well on the Holmden Farm comes in at 650 barrels a day. This strike launches the city of Pithole, Pennsylvania, which will have sixteen thousand residents by summer.

On February 21, Col. E.A.L. Roberts demonstrates his process of torpedoing on The Ladies Well outside of Titusville, Pennsylvania.

On April 9, the surrender of Southern armies at the Appomattox Court House signals the end of the American Civil War.

On April 14, John Wilkes Booth assassinates President Abraham Lincoln. Six weeks later, the Dramatic Oil Company's first major well comes in. Booth had a share in the company.

In September, the first shipment of oil in a railroad tank car arrives in New York. The car is constructed by placing two wooden stock tanks on a flat rail car.

On October 10, the first oil pipeline is finished by Samuel Van Syckle. It is a two-inch pipe that connects The US Well at Pithole with Miller Farm on the Oil Creek Railroad.

1866   Col. E.A.L. Roberts is granted a patent for his concept of the exploding torpedo and fracing oil wells. He was involved in two thousand lawsuits regarding this patent, and it is said that he never lost a lawsuit.

1867   On July 15, experiments are conducted to see if oil could be used as a fuel for steam locomotives.

1868	Robbers steal $250,000 from a safe at the Benninghoff Farm.
1869	The exploration and drilling for oil begins to move out from Venango County into Armstrong, Butler and Clarion Counties at first, and then beyond to the rest of the world.
	A process for deodorizing lubricants made from petroleum is discovered by Joshua Merrill.
1870	John D. Rockefeller organizes the Standard Oil Company as a corporation in Ohio.
1871	The Oil Exchange is formed in Titusville, Pennsylvania.
	The boroughs of Oil City and Venango incorporate to become Oil City, Pennsylvania.
1872	For domestic uses, natural gas begins to be pumped to Titusville, Pennsylvania.
1873	The Commonwealth of Pennsylvania grants Edwin L. Drake an annual pension of $1,500.
1874	The oil field in Bradford, Pennsylvania, booms.
	The Oil City Exchange is organized in Oil City.
1875	The first commercial oil well is drilled in California.
1876	The Oil City Gas Company is charted in Oil City.
1878	The Standard Oil Company controls 90 percent of the refining capacity in the United States.
	The *Evening News* in started in Franklin, Pennsylvania.

1879	The first pipeline of over a hundred miles, the Tidewater, is completed from Williamsport, Pennsylvania, to Bradford over the Allegheny Mountains.
1880	On September 14, the Oil Exchange opens in Oil City.
	On November 8, at the age of sixty-one, Edwin L. Drake dies in Bethlehem, Pennsylvania.
1881	The National Transit Company is organized by the Standard Oil Company.
	The Oil City Telephone Exchange is built in Oil City.
1882	John D. Rockefeller, along with others, organizes the Standard Oil Trust.
1883	The first oil well drilled in Wyoming is completed.
	The Great High Water Flood hits Oil City.
1884	The Producer's Petroleum Exchange is organized.
1890	The Sherman Antitrust Act is passed by the United States Congress.
1892	A seventy-two-foot-high steel derrick is made available through catalog order; a first for the industry.
	A formal resolution to dissolve is passed by the board of the Standard Oil Trust.
1893	For the first time, kerosene is pumped by pipeline from the Pennsylvania Fields to Wilkes-Barre, Pennsylvania, a distance of 252 miles.
1895	California produces 1.2 million barrels of crude oil.

1899   Standard Oil is reorganized as a holding company in New
               Jersey.

1901   On January 10, Spindletop, also known as the Lucas
               Gusher, comes in near Beaumont, Texas, signaling the
               beginning of the shift in the industry from the Pennsylvania
               Fields to elsewhere.

GLOSSARY OF PETROLEUM
INDUSTRY LANGUAGE

What follows here is a selection of some of the words and terms that were developed and used in the Oil Region. While they are defined, keep in mind that the definitions used here refer to Pennsylvania's Oil Region. Therefore, the meaning of the terms may have shifted somewhat in other regions. Still, this list should be enough to enable you to begin to talk like a real oilman.

Atlantic Ocean: A well producing salt water and little or no oil.

attic hand: A worker in the oil fields who usually works high up in the rigs and on the top of tanks.

baldheaded: One way of describing a drill bit on which the edge has worn off.

band wheel: The wheel in the belt house around which the band ran. It was used to transfer the energy produced by the engine to the drilling rig.

barefooted well: A well that was drilled without the use of casing or screen.

barker: An extension placed on the end of an engine exhaust pipe to change the sound it makes and differentiate it from other engines.

barrelhouse: A drinking establishment where the bartenders were known to put drunken customers into empty barrels and roll them out the door and into the street.

biscuit cutter: A jesting putdown of one driller by another. The term comes from the biscuit-shaped marks left on the main sill when pipes are dropped on it end first.

black gold: Another name for petroleum. This term probably stems from the fact that the early oil industry coincided with the California and Yukon Gold Rushes.

boomer: An oil field worker who followed the booms from field to field as they happened. A boomer would often give up a good job with a good company earning good wages in order to head off for the next oil boom.

bull wheel: The large wheel on the drilling rig around which the drilling cables were wrapped and from which they played out.

busted: A well, a business or a man from whom nothing more can be pumped, promoted or persuaded due to a lack of flowing material at the bottom of the hole.

calf wheel: The smaller wheels set near to where the driller stood at the throttle that were used to pull pipe in or out of the well and to lower pipe down the well itself.

Carbon Oil: One of the names used for petroleum leading up to and during the early petroleum age.

cash on the barrelhead: A term to mean that one has or requires hard cash to complete the transaction. This term is said to have originated in the oil fields where the end of a barrel was often the closest flat surface to work on.

Christmas tree: An assemblage of pipes, valves and other fittings placed on the top of a well before the pipeline in order to control and manage the flow of oil or gas coming from the well.

crude/crude oil: Petroleum as it comes out of the ground, to which nothing has been done other than pumping it up from the depths.

crude skinners: An early name for the teamsters who transported oil in barrels from wellhead to railhead or boat dock.

Curbstone Exchange: The sidewalk in front of the offices of Lockhart, Frank and Company in Oil City, Pennsylvania, during the 1870s. It was here where oilmen gathered to discuss the news of the day, exchange information and buy and sell oil.

dead-in-a-hurry: A name used to describe those who transported and worked with nitroglycerin. It came into usage because of the nature of the work being so dangerous that many insurance companies would not cover those who chose to do the work.

derrick apples: A term used to describe the things that have fallen to the main sill under the derrick. These items could be anything from nuts and bolts to mud to pieces of rope to anything else that happened to fall on the main sill.

driller/rope choker: The man charged with drilling the well. This person was the boss at the wellhead and ran everything. Also called a rope choker due to the fact that he often had his hand wrapped around the

A drill house and rig tower from the Oil Region.

drilling rope in order to keep track of what was happening at the end of the tool string in the bottom of the well.

farmer's sand: This was an inside joke used when a dry well was drilled. The well was dry because it was not drilled down into the farmer's sand. If anyone asked the depth of the farmer's sand, it was always so deep that no one could drill into it, but that was where the best oil lay.

freshet: An artificial flood that is used to increase the depth of a creek or river for a short time. This is done by creating small ponds along the run of the creek or river using gated dams. On the posted day and at the designated hour these gates would be opened, allowing the water behind them to flow out and cause the flood. Originally developed by the lumber industry, this process was adopted by the oil industry so that petroleum could be shipped year-round, regardless of the level of the creek or river.

go-devil: A small ring-shaped device that was placed around the cable and was used to lower a torpedo of nitroglycerin down an oil well. This device was dropped down the cable to cause the nitroglycerin to explode. It was called a go-devil because the shooters were told that when they released it, they better go like the devil himself was chasing them.

greasers: The name for the men who worked in the oil fields, as well as the clothing they wore. The clothing normally consisted of high-topped boots, corduroy or denim trousers, a flannel shirt and a hat speckled with slush.

Guiper: A flat-bottomed boat of between fifteen to twenty-five feet in length and capable of holding between twenty-five to fifty barrels of oil. It was used for transporting oil from the wellhead downstream to the refineries.

gusher/flowing well: A well containing both oil and natural gas under pressure. It was the pressure of the natural gas that caused the oil to flow or gush to the surface, often throwing it high in the air and causing it to rain black gold all around the wellhead.

headache post: A post that is approximately six inches by eight inches and about ten feet long. It is set upright on the derrick floor as the well is being drilled, placed on the main sill on the rig so that it is under the beam as it rises up and down. The main purpose of the headache post is to stop the beam from falling on someone if anything were to break. It was also used as a gathering place at the wellhead.

hogshead: A measure of liquid that varies from person to person and location to location. Normally it varies between a minimum of 63 gallons (or two barrels) to a maximum of 140 gallons (or four barrels). It was later standardized to the lower numbers.

ile: An early way to pronounce oil.

independent: A term that came into usage with the monopoly of Standard Oil. It was used to refer to any business that worked outside of the Standard Oil Conglomeration.

kicking down a well: The process of using a spring pole to drill a well. The spring pole is bent over the desired wellhead and a rope with a loop at the end of it is attached to the pole. One then put a foot into the loop and began kicking downward in order to move the drill string up and down.

Kosmos Burner: A device for testing the candlepower of illumination petroleum and other oils.

lease: An agreement between the landowner and the production company to allow the production company to drill on the landowner's property. The normal terms of a lease were for a small amount of money on the front end and a part of the profits when the well came in.

lease hound: A man who made his living buying and selling oil leases.

mystery well: A well drilled in complete secrecy, or as much secrecy as possible. The idea was to keep the details of the well away from the rest of the industry until such time as it came in.

oil drummer: A person who traveled the countryside shilling petroleum products to consumers.

oil fever: That peculiar disease that overcomes people when they get the idea that they can make their fortune with the next strike.

oil field dove: A lady of the evening who followed the oilmen from boom to boom and field to field.

oil payment: A complex financial transaction that evolved on the oil fields and made many a man rich. It worked like this: a fellar we'll call Paul owned a lease to drill on another fellar's land. Let's call this second fellar Peter. In order for Paul to drill on Peter's land, he cuts a deal with Thomas to drill the well. Now, as we said, Paul has no money, so he agrees to pay Thomas a set amount out of the profits from the first oil produced on Peter's land. This amount was usually figured as a percentage per barrel, and while it varied from agreement to agreement, it was called an oil payment. The thing to remember is that while an oil payment was a risk, it was still worth something, and if Thomas found himself in need of some ready cash, he might just sell his oil payment to someone else. It was in this buying and selling of oil payments that many men got rich.

oil scout: A person who makes his living by keeping track of the new wells, fresh strikes and other information from the oil fields and then reporting this information back to the major companies who employed them.

pencil pusher: A name earned by oil scouts due to their habit of coming around oil wells with a pad and pencil in hand, asking all sorts of prying questions in order to have something to report back to the major companies who employed them.

pulling a well: The act of removing all the rods, pipes and tubing from a well in order to clean it out and get it producing again.

roughneck: In general, anyone who works in the oil field. It is said that this term stems from the fact that anyone who chooses to work in the oil fields needs a rough neck in order to be able to do so. Eventually, this term was used solely for the person who did what he was told while in charge of the pipe above the ground. He looked after the pipe, breaking it out of the stack as it was needed or cleaning and stacking the pipe when a well was pulled.

roustabout: A general term for anyone who is working in the oil fields doing whatever needs to be done. Usually this is a person without much training who is just starting out. Among his duties was looking after the wells, the lines and the tanks. He was in charge of making sure that everything was as it should be and of picking up the broken rods, junk pipe and assorted whatnots that littered the oil field and made it a dangerous place to work.

shooter/oil well shooter: Originally someone who worked for the Roberts Torpedo Company shooting oil wells. Later, the term applied to anyone who shot wells.

shooting a well: The process developed by Col. E.A.L. Roberts to increase the production rate of an ailing oil well. In short, a tube containing nitroglycerin was lowered into the well and then exploded, causing new fractures to form in the rock and sand and allowing the oil to flow into the well.

soiled dove: A lady of the evening who usually worked for one of the bars or parlors in the boomtowns.

spring pole: A length of pole or tree approximately ten to fifteen feet long that is flexed over a well and used as the return spring when kicking down a well.

tool dresser/toolie: Second in command to the driller at the wellhead, the toolie's main job consisted of keeping the drilling tools sharp and ready for the driller.

torpedo: A long tube that was used to hold the nitroglycerin that was lowered down a well in order to fracture or "frac" the ground around the bottom of the well and hopefully increase the production of the well.

FURTHER READING

For those of you who have a more scholarly bent, I am placing here a short bibliography of other books you may want to take a look at. Some of these books are in print and some of them are long out of print. As such, you will have to go out and find them for yourself. I do have quite a few of them in my own personal collection and have referred to them many a time for both pleasure and business.

American Petroleum Quarterly, Centennial Issue. New York, 1959.

Ball, Max Waite. *This Fascinating Oil Business*. New York: Bobbs-Merrill Company, 1940.

Beaton, Kendall. *Enterprise in Oil: A History of Shell in the United States*. New York: Appleton-Century-Crofts, 1957.

Boatright, Mody Coggin. *Folklore of the Oil Industry*. Dallas: Southern Methodist University Press, 1963.

————. *Gib Morgan: The Minstrel of the Oil Fields*. Dallas: University of North Texas Press, 1945.

Boatright, Mody Coggin, and William Owens. *Tales from the Derrick Floor*. Garden City, NY: Doubleday, 1970.

Bone, J.H.A. *Petroleum and Petroleum Wells*. Philadelphia: J.B. Lippincott, 1865.

Boone, Lalia. *Petroleum Dictionary*. Norman: University of Oklahoma Press, 1952.

Carll, John Franklin. *The Geology of the Oil Region of Warren, Venango, Clarion, and Butler Counties*. Harrisburg, PA: Board of Commissioners for the Second Geological Survey, 1880.

Clark, James A., and Michael T. Halbouty. *Spindletop*. New York: Random House, 1952.

Clark, J. Stanley. *The Oil Century: From the Drake Well to the Conservation Era.* Norman: University of Oklahoma Press, 1958.

Cone, Andrew, and Walter R. Johns. *Petrolia: A Brief History of the Pennsylvania Petroleum Region, Its Development, Growth, Resources, etc., from 1859 to 1869.* New York: D. Appleton, 1870.

Darrah, William C. *Pithole: The Vanished City.* Self-published, 1972.

Dolson, Hildegarde. *The Great Oildorado: The Gaudy and Turbulent Years of the First Oil Rush.* New York: Random House, 1959.

Eaton, S.J.M. *Petroleum: A History of the Oil Region of Venango County, Pennsylvania, Its Resources, Mode of Development, and Value; Embracing a Discussion of Ancient Oil Operations.* Philadelphia: J.P. Skelly and Company, 1866.

Fanning, Leonard M. *The Rise of American Oil.* New York: Harper, 1948.

Forbes, Gerald. *Flush Production: The Epic of Oil in the Gulf-Southwest.* Norman: University of Oklahoma Press, 1942.

Gibb, George Sweet, and Evelyn H. Knowlton. *The Resurgent Years, 1911–1927.* New York: Harper, 1956.

Giddens, Paul Henry. *The Birth of the Oil Industry.* New York: Macmillan Company, 1938.

———. *Early Days of Oil.* Princeton, NJ: Princeton University Press, 1948.

———. *Standard Oil Company (Indiana): Oil Pioneer of the Middle West.* New York: Appleton-Century-Crofts, 1955.

Glasscock, C.B. *Then Oil Came.* Indianapolis: Bobbs-Merrill Compay, 1938.

Henry, J.T. *The Early and Later History of Petroleum with Authentic Facts in Regard to its Development in Western Pennsylvania.* Philadelphia: J.B. Rodgers Company, 1873.

Hildy, Ralph Willard, and Murial E. Hildy. *Pioneering in Big Business, 1882–1911.* New York: Harper, 1955.

Hitchcock, Roxanne. *Lube Lingo.* Oil City, PA: Oil Region Books, 1999.

House, Boyce. *Oil Boom: The Story of Spindletop, Burkburnett, Mexia, Smackover, Desdemona, Ranger and the Wildcatters who Found Them.* Caldwell, ID: Caxton Printers, 1941.

James, Marquis. *The Texaco Story: The First Fifty Years.* New York: The Company, 1953.

Knowles, Ruth Sheldon. *The Greatest Gamblers: The Epic of American Oil Exploration.* New York: McGraw-Hill, 1959.

Larson, Henrietta M., and Kenneth W. Parker. *History of Humble Oil and Refining Company: A Study in Industrial Growth.* New York: Harper, 1959.

McElwee, Neil. *Oil Creek...The Beginning: A History and Guide to the Early Oil Industry in Pennsylvania.* Oil City, PA: Oil Creek Press, 2001.

————. *Standard Oil Co. Men in the Early Oil Region*. Oil City, PA: Oil Creek Press, 2006.

McLaurin, John J. *Sketches in Crude Oil: Some Accidents and Incidents of the Petroleum Development in All Parts of the Globe*. Harrisburg: self-published, 1896.

Michener, Carolee, and the Venango County Historical Society. *Oil, Oil, Oil!* Franklin, PA: Seneca Printing, 1997.

Munn, S.W. *Useful Information for Oil Men*. Mannington, WV: Press of the Enterprise, 1900.

Pratt, Wallace. *Oil in the Earth*. Lawrence: University of Kansas Press, 1942.

Rister, Carl Coke. *Oil! Titan of the Southwest*. Norman: University of Oklahoma Press, 1949.

Smiley, Alfred Wilson. *A Few Scraps, Oily and Otherwise*. Oil City, PA: Derrick Publishing Company, 1907.

Steele, J.W. *Coal Oil Johnny: Story of His Career as Told By Himself*. Franklin, PA: self-published, 1902.

Tait, Samuel W., Jr. *The Wild Catters*. Princeton, NJ: Princeton University Press, 1946.

Taylor, Frank J. *The Black Bonanza: How the Oil Hunt Grew into the Union Oil Company of California*. New York: Whittlesley Company, 1956.

Whiteshot, Charles. *The Oil Well Driller*. Mannington, WV: C.A. Whiteshot, 1905.

Visit us at
www.historypress.net

www.ingramcontent.com/pod-product-compliance
Lightning Source LLC
Chambersburg PA
CBHW070349100426
42812CB00005B/1469